Regulation of Scientific Inquiry

AAAS Selected Symposia Series

Regulation of Scientific Inquiry

Societal Concerns with Research

Edited by Keith M. Wulff

Routledge
Taylor & Francis Group

LONDON AND NEW YORK

First published 1979 by Westview Press

Published 2019 by Routledge
52 Vanderbilt Avenue, New York, NY 10017
2 Park Square, Milton Park, Abingdon, Oxon OX14 4RN

Routledge is an imprint of the Taylor & Francis Group, an informa business

Library of Congress Catalog Card Number: 78-24738

ISBN 13: 978-0-367-28552-4 (hbk)
ISBN 13: 978-0-367-30098-2 (pbk)

About the Book

The increase in regulations affecting the conduct of scientific research, and the debate about their appropriateness and effectiveness, reflect societal concerns with fundamental questions raised by certain types of scientific inquiry. This book addresses issues of ethics, accountability, and conflict as they relate to the rights of inquiry, the rights of citizens, and the role of government in a research-oriented society.

About the Series

The *AAAS Selected Symposia Series* was begun in 1977 to
provide a means for more permanently recording and more
widely disseminating some of the valuable material which is
discussed at the AAAS Annual National Meetings. The volumes
in this *Series* are based on symposia held at the Meetings
which address topics of current and continuing significance,
both within and among the sciences, and in the areas in which
science and technology impact on public policy. The *Series*
format is designed to provide for rapid dissemination of
information, so the papers are not typeset but are reproduced
directly from the camera-copy submitted by the authors, without copy editing. The papers are organized and edited by
the symposium arrangers who then become the editors of the
various volumes. Most papers published in this *Series* are
original contributions which have not been previously published, although in some cases additional papers from other
sources have been added by an editor to provide a more comprehensive view of a particular topic. Symposia may be reports of new research or reviews of established work, particularly work of an interdisciplinary nature, since the AAAS
Annual Meetings typically embrace the full range of the
sciences and their societal implications.

WILLIAM D. CAREY
Executive Officer
American Association for
the Advancement of Science

Contents

About the Editor and Authors

Keith M. Wulff, assistant professor of sociology in the Department of Sociology and Social Work at Concordia College, is interested in the sociology of science and of religion. His research has focused on the attitudes of the public toward science and educational institutions, and he has recently been awarded a grant to plan a Tri-College (North Dakota State University, Moorhead State University, and Concordia College) Social Science Research Institute.

Kurt W. Back, James B. Duke Professor and chairman of the Department of Sociology at Duke University, has extensive experience in social psychology, gerontology, and medical sociology. He has received two Special Research Fellowships from the National Institutes of Health and is a fellow of the American Sociological Association. He is the author of many articles on language, society and subjective experience, self-disclosure and the volunteer, self-help groups, life graphs and life events, and other subjects, and his books include The June Bug: A Study in Hysterical Contagion *(with A. Kerckhoff; Appleton, Century, Crofts, 1968) and* Beyond Words: The Story of Sensitivity Training and the Encounter Movement *(Russell Sage Foundation, 1972).*

Barry M. Casper, professor of physics and chairman of the Physics Department at Carleton College, specializes in theoretical physics. He was chairman of the American Physical Society's Forum on Physics and Society, served on the National Council of the Federation of American Scientists, and was a candidate for Lt.-Governor of Minnesota in 1978. He has published articles on scientists and the Congress, Presidential science advisors, the "science court" proposal, laser enrichment, and technology assessment, and is coauthor of Revolutions in Physics *(with R.J. Noer; W.W. Norton, 1972). He is currently working on a book entitled* Scientists and the Politics of Technology.

Thomas I. Emerson, Lines Professor of Law Emeritus at Yale University, specializes in constitutional law. Among his publications are three books: Toward a General Theory of the First Amendment *(Random House, 1966),* The System of Freedom of Expression *(Random House, 1970), and* Political and Civil Rights in the United States, 4th ed. *(with D. Haber and N. Dorsen; Little, Brown, 1976).*

Eliot Freidson, professor of sociology at New York University, has written and edited books on the organization of health care, the role of professions in the modern world, the responses of physicians to prepaid group practice, and other topics. He has also published articles on testimonial privilege for researchers and problems of confidentiality in research. He received the Sorokin Award of the American Sociological Association in 1972 and is a member of the Institute of Medicine of the National Academy of Sciences.

Harold P. Green is professor of law and director of the Law, Science, and Technology Program at The George Washington University National Law Center. He is founding fellow and member of the board of directors of the Institute of Society, Ethics, and the Life Sciences of the Hastings Center, a member of the panel on Efficiency in Environmental Decision-Making appointed by the Nuclear Regulatory Commission, a member of the Biohazards Committee of the Frederick National Cancer Research Center, and consultant to the Director of the National Institutes of Health with respect to legal problems involved in recombinant DNA molecule technology and to Batelle Northwest Laboratories concerning preparation of materials for ERDA's Environmental Impact Statement on nuclear waste management. He has testified before various legislative committees on these and related issues.

Philip Handler, president of the National Academy of Sciences since 1969, was previously James B. Duke Professor of Biochemistry at Duke University. He has been president of the American Society of Biological Chemists, chairman of the National Science Foundation's National Science Board, chairman of the board of the Federation of American Societies for Experimental Biology, and a member of the President's Science Advisory Committee (1964-1968 and 1969-1972) and of the Rockefeller University Board of Trustees. He has been active in numerous professional organizations, has received 22 honorary degrees and many other honors and awards, and has published over 200 papers and books.

Andre E. Hellegers, director of the Kennedy Institute of Ethics at Georgetown University, is also professor of obstetrics-gynecology and of physiology-biophysics at Georgetown.

He has served on the boards and committees of numerous professional organizations, on the editorial boards of several medical journals and one theology journal, and is past president of the Society for Gynecologic Investigation and of the Perinatal Research Society. Awarded the Linacre Award in 1974 and an honorary D.H.L. from Wheeling College in 1975, he has published some 180 articles in professional journals.

Hans Jonas, Alvin Johnson Professor Emeritus of philosophy of the graduate faculty of the New School for Social Research, is currently interested in the philosophy and ethics of technology. He has studied and written about the history of philosophy and religion, gnosticism, and the philosophy of organism. He was honorary president of the International Colloquium on Gnosticism and is a fellow of the American Academy of Arts and Sciences and of the Institute of Society, Ethics and the Life Sciences, from which he received the Henry Knowles Beecher Award (1978). He has given testimony on regulation of biomedical research and has recently published several articles on recombinant DNA. His latest book is **Faith, Reason, and Responsibility** *(Scholars Press/ Harper and Row, in press).*

Hans O. Mauksch, professor of sociology and of family and community medicine at the University of Missouri-Columbia, has studied the Sociology of health care, health care institutions and occupations, and the sociology of teaching. He served as executive officer of the American Sociological Association (1975-1977) and has published on socialization and career choices in medicine and nursing, patient care, hospital organization, family health, teaching, and conditions of research.

David J. Newburger, assistant professor of law at Washington University School of Law, specializes in the regulation of business and is a member of The District of Columbia and Supreme Court Bars. He has testified before the U.S. Senate and House of Representatives on recombinant DNA research regulation, and is coauthor of a literature study entitled **The Effects of Regulation on Technological Innovation in the Chemical and Allied Products Industries** *(NSF, 1975). He is also coauthor of* **Regulation and Market Price: Effects on Process Innovation—The Case of the Ammonia Industry** *(Westview, in press).*

Gerard Piel is president and publisher of **Scientific American**. *He is a fellow of the American Academy of Arts and Sciences, an overseer of Harvard College, and a trustee of numerous educational institutions, museums, and foundations. He has been awarded several honorary degrees and is the* *author of* Science in the Cause of Man *(Knopf, 1961) and* The Acceleration of History *(Knopf, 1972).*

Albert J. Reiss, Jr., William Graham Sumner Professor of Sociology at Yale University, has conducted research on deviant behavior and social control, sociology of law, and methodology. He is a member of the Panel on Research on Effects of Deterrence and Incapacitation of the National Academy of Sciences and chairman of the Social Science Research Council Advisory and Planning Committee on Social Indicators. His publications include The Police and the Public *(Yale University Press, 1971), "Selected Issues in Informed Consent and Confidentiality with Special Reference to Behavioral Science Inquiry: A Report to the National Commission on the Protection of Human Subjects in Biomedical and Behavioral Research" (1976).*

Lee N. Robins, professor of sociology and psychiatry at Washington University School of Medicine, specializes in psychiatric sociology, social deviance phenomena, and epidemiology. She is a member of the National Advisory Council on Drug Abuse, and she received the 1978 Paul Hoch Award of the American Psychopathological Association, the 1978 Pacesetter Research Award of the National Institute on Drug Abuse, and the Research Scientist Award of the National Institute of Mental Health since 1970. She has published numerous articles, books, and monographs, including Life History Research in Psychopathology, Volume 2 *(University of Minnesota, 1972) and* A Follow-Up of Vietnam Drug Users *(Executive Office of the President, 1973).*

Robert L. Sinsheimer, chancellor of the University of California at Santa Cruz, is a biophysicist and molecular biologist. He is past president of the Biophysical Society and a member of the National Academy of Sciences and of the American Academy of Arts and Sciences, and has received honorary Sc.D.'s from St. Olaf University and Northwestern University. He was named California "Scientist of the Year" in 1968 and was awarded the Beijerinck Virology Medal from the Royal Netherlands Academy in 1969. He is the author of over 200 articles on the molecular biology of DNA and viruses and on various implications of science.

Adlai Stevenson, U.S. Senator from Illinois since 1970, is chairman of the Subcommittee on Science and Space of the Committee on Commerce, Science, and Transportation, which exercises jurisdiction over the federal government's programs in science and technology. He is also chairman of the Ethics Committee, which insures compliance with the Senate's Code of Ethics.

Dael Wolfle, professor emeritus at the University of Washington Graduate School of Public Affairs, was executive officer of the American Association for the Advancement of Science and publisher of Science *magazine from 1954 to 1970. He has served on many boards and committees of public and private agencies concerned with research and development and with scientific and other specialized personnel, and has published numerous articles and books, including* Science and Public Policy *(Univ. of Nebraska, 1959) and* The State of Academic Science: Background Papers *(coeditor;* Change Magazine Press, *1978).*

Introduction

Keith M. Wulff

The increase in the regulations affecting the conduct of scientific inquiry in recent years has prompted, from scientists and the general public, responses ranging from gloom to "it is about time." The AAAS held a symposium at its February 1978 annual meeting to examine the appropriateness and effectiveness of the regulations affecting scientific inquiry.[1] The question making this examination a fundamental issue was, What is the function of government in balancing the rights of inquiry with the rights of citizen to privacy and protection from harm?

The papers given at that symposium form the core of this book. Additional articles which cover points not covered at the symposium or which enlarge on what was covered are also included. Jonas's, Piel's, and Sinsheimer's papers were included because they enlarge on points made by the other authors and their work provides the background for some of the other authors' comments.[2] The papers of Stevenson, Handler, Newburger, Emerson, and one of Green's were originally given at the congressional hearings on recombinant DNA regulations and cover material not covered at the symposium.[3] The other paper of Green's is from the September–October 1978 Bulletin of Atomic Scientists.[4]

After a debate has been carried on for a period of time the central issues under debate sometimes become lost in the rhetoric. The papers in the first section--chapters 1-7--were selected because they deal with the basic issues. Hellegers makes a distinction between imposition on freedom of inquiry and the government not funding certain research. Freedom of inquiry does not mean others have to pay for the research. Hellegers thinks the public should be represented in the decision making process of funding research. These decisions should not be made only by a community of scientific peers.

The arguments for or against limitation of research can become very abstract. It is necessary at some point for the arguments to be specifically worded for they must be applied to specific cases. Casper is very specific. He first distinguishes between limitation of research based on deleterious consequences of the inquiry itself and limitations based on potential harm resulting from the application of that research. Using the emerging technology of laser enrichment of uranium as an example he argues for the early restriction of research based on potential harm resulting from the application of the new knowledge. His argument for early restriction of inquiry is that after a certain point the technology connected to the new knowledge takes on a life of its own. At that point it is exceedingly difficult to stop.

Casper and Hellegers both agree that the public needs to be involved in determining research priorities. In the chapter containing the discussion of their presentation they enlarge on how the public should become involved. The feasibility and desirability of technological assessment is also discussed. Hellegers enlarges on his idea of using funding as a means of regulating research. Other practical questions concerning regulation of research are raised by the symposium audience.

Hans Jonas's article is an excellent statement concerning the accountability of science as an agent of social action. Jonas says that the change in the way science is done has resulted in a lapse of scientific freedom in experimentation. He shows how science has moved from theory to action. Society does not give the same freedom to action as it does to thought.

The Hastings Center Report published edited reactions to Jonas's comments by Piel and Sinsheimer.[4] Rather than reprint those articles we have included more detailed works by the same authors. Piel's article is a very strong defense of continued freedom of inquiry for science. He sees the tagging of public funding of science for certain projects as narrowing the development of our country's scientific talent. Piel does not agree with the people who refer questions of truth to scientists and questions of value to others. He believes scientists are the best qualified members of the public to frame science policy.

Many individuals find it distasteful to think about limiting inquiry or to think that new knowledge may bring

about more harm than good. Sinsheimer suggests we should
think about these questions. He suggests that all knowledge
may not necessarily be good and that the single-minded pursuit of new knowledge could destabilize society so that
society would no longer support the scientific enterprise.
Back, while not supporting Sinsheimer, points out that the
move toward unrestrained pursuit of science is a novel development in human history. The questions Sinsheimer asks are
questions the public has always been willing to ask of those
pursuing knowledge. Back shows the limits individuals put on
knowledge they think others should have of them and knowledge individuals feel would be dangerous if they possessed
it themselves. In general, people are more likely to tolerate invasion of their privacy for the benefit of science if
they think they have something to gain out of the experience
than if it is just for acquisition of more knowledge.

Back may appear to contradict the other authors who
stress the freedom that scientists have had in the past.
There is no contradiction for the other authors define as the
past what Back sees as recent. He defines recent times as
beginning with the renaissance and emphasizes how freedom of
inquiry has grown since then. Back provides a needed
historical perspective to the problem.

Recently much of the discussion concerning regulation
of research has dealt with recombinant DNA research. The
authors in the second section of the book look at the issues
surrounding the regulation of this type of research. In
looking at this case history we can clearly see some of the
problems that must be dealt with in the regulation of
research in general. Senator Adlai Stevenson's remarks
present a short history of the DNA controversy and the considerations he thinks Congress should take into account when
considering legislation to regulate scientific inquiry.
Handler's comments are very specific. He points out the
importance of recombinant DNA research and states why he
thinks this type of research can be carried out safely.

Green's article differs from the previous two in that
he looks at how government might assess benefits and risks
in scientific research such as recombinant DNA work and to
what extent government should regulate privately funded
research. Green points out several parallel situations
between regulation of atomic energy research and recombinant
DNA.

Most of the authors of previous articles mentioned the
legal right under the first amendment for scientists to do
research without regulation by the government. The authors

of the three articles in the third section--chapters 11-13
--discuss this and other legal questions. Emerson and Green
examine specifically whether the rights of free speech apply
to scientific research. They also discuss whether an appeal
to the first amendment is the best procedure to prevent
regulation of scientific inquiry. Newburger's article is
somewhat broader. He looks at how legal regulations can be
used to regulate scientific research and if that is the best
means of regulating research.

The points raised by Newburger are some of the most
crucial, but they are often overlooked during discussion of
regulations. He states conditions that must be met if
research is to be successfully regulated by prohibiting
certain conduct. The need for flexibility in enforcing
regulations is probably seen by most people. However, the
effect on research and development of the uncertainity concerning what is or is not acceptable behavior may not be as
readily seen. Newburger discusses this tension. He also
points to the need of distinguishing between academic and
commercial activities.

Several of the authors point out that the regulation of
research is not new. What is new is that certain disciplines
are being regulated for the first time. One question is
whether a model for regulating one discipline can be applied
to a different discipline. In the fourth section--chapters
14-17--the authors look at this question and at the possible
effect government regulations may have on social sciences.
Robins shows how the important work she has done in the past
would not be possible under the guidelines proposed by the
Privacy Protection Study Commission. Reiss shows how the
Bio-medical model for regulation of research does not apply
to the social sciences. Mauksch points out problems that
arise when the local board regulating research is made up of
members from a discipline other than the one being regulated.
He also points out that regulations have different effects
on the different social classes.

Freidson in his article takes a view opposing Robins,
Reiss, and Mauksch. Freidson agrees that government regulations have made research more difficult but thinks the
positive features of regulation out-number the negative.
Freidson succinctly states several important issues. One of
these issues is the potential misuse of social science data
by the government and the need to protect against that.

The last two articles summarize the book and put the
various articles into critical perspective. Wolfle discusses
the papers that were given at the 1978 AAAS symposium. He

clearly points out what progress he believes has been made
and what still needs to be done if we are to have a workable
and fair method of regulating research. Wulff stresses the
responsibility members of the scientific community have in
regulating their own research. He points out that attacking
regulations in principle as impinging on a scientist's freedom may not help the scientist. What would be more fruitful
would be to stress the impact that specific regulations would
have on innovation and the growth of knowledge.

References

1. This symposium was arranged by Hans O. Mauksch, Rosemary
   A. Chalk and Keith M. Wulff. Kathryn Wolff and Jill
   Storey have been very helpful in editing this book. Its
   completion was due largely to the help of Rosemary A.
   Chalk, Carol Buck, Annette Larson, Jo Engelhardt, and
   Judith Livdahl Wulff.
2. Hans Jonas, Hastings Center Report, 6 (1976): 15-17;
   Robert L. Sinsheimer, Daedalus, 107 (1978): 23-35.
   Both articles reprinted by permission of authors and
   publishers.
3. Adlai Stevenson, Congressional Record, Sept. 22, 1977,
   pp. S15410-15413; Philip Handler, Com Hearings, Senate
   Com. on Commerce, Science and Transportation, Nov. 2,
   1977; Harold P. Green and Thomas I. Emerson, Com.
   Hearings, Senate Com. on Science and Transportation,
   Nov. 10, 1977; Green's article, "The Boundaries of
   Scientific Freedom," has also appeared in Harvard
   University Newsletter on Science, Technology and
   Human Values, 20 (1977): 17-21. Reprinted by
   permission of publisher.
4. "The Recombinant DNA Controversy: A Model of Public
   Influence," reprinted by permission of the Bulletin
   of the Atomic Scientists. Copyright (c) 1978 by the
   Educational Foundation for Nuclear Science.
5. Gerard Piel, Hastings Center Report, 6 (1976): 18-19;
   Robert L. Sinsheimer, Hastings Center Report, 6
   (1976): 18.

Part I
Regulation of Scientific Inquiry

1

The Ethical Dilemmas
of Medical Research

Andre E. Hellegers

From the program of this meeting it is clear to me that
I should not restrict my comments today to dilemmas in medical research alone, for this program is about scientific
research in a much broader context than medicine alone. As
a consequence I propose simply to use medical research as a
paradigm in a much broader setting. What I perceive this
program to be about is the subject of freedom of scientific
inquiry in general and I see myself as charged to approach
the broader subject from the narrower perspective.

It seems to me to be odd that we should hear mumbles,
if not cries, of anguish from some quarters in science, as
if freedom of inquiry is in peril in these United States.
To me that is an absurd posture. Never in the history of
mankind has any research endeavor been as royally funded as
has the endeavor of scientific inquiry in the U.S. If we
take only the funds available for scientific inquiry through
the medical establishment, the physical sciences establishment and the military establishment, the generation of
scientists preceeding us would find them mind boggling.

It is of course a fact that research and development
money is no longer subject to those automatic increases of
15 years ago and it may be true that it may only barely be
keeping pace with the inflation rate, but what endeavors do
better? It is surely unseemly that we should expect to be
supported at a constantly increasing rate, sight unseen,
when all other aspects of public expenditures are under increasing scrutiny, to the point of insisting on zero based
budgeting. I sometimes think we are assuming the posture of
the medieval clergy, who would insist that they should not
be scrutinized because their product was eternal life and
truth (albeit in the hereafter). Similarly, some insist
that we are the custodians of eternal life and verities (in

9

the here and now), so that we should also be immune to inquiry, while we insist on our own freedom of inquiry.

I draw this parallel between religion formerly and science today, because I was struck by a recent comment made by Gerard Piel, publisher of the <u>Scientific American</u>. In a comment on an article by Hans Jonas on freedom of scientific inquiry and the public interest, in the <u>Hastings Center Report</u> of August, 1976, Mr. Piel asserted that the peer review system is a mechanism to ensure a democratic process in science. He holds that this system makes it possible for the enormous financial power of the federal government and all the other kinds of authority attached to it, to be decoupled from the support of science. He states that "the peer review group removes an authority external to science from the decisions that are made about its ongoing work." He calls this peer review system a "new invention in self government." It is not clear to me how Mr. Piel defines the term "peer review." Does he mean only fellow scientists or does he mean all citizens? I read him to mean the former, while I hold government still to be the most representative body of the citizenry in a democracy.

If I read him aright I am struck by the parallels with religious notions of the past. I am reminded of the days when the clergy could only be judged by clerical courts and not by civil ones. This notion that only the clergy can judge clergy, only the military the military and only scientists the scientists, strikes me as a notion which can only harm these enterprises. About a decade ago, the then Senator, now Vice President, Mondale suggested the establishment of a commission to study the impact of medical science on people. He was strongly supported by witnesses, mainly lawyers, who believed such a study commission necessary to protect patients. He was opposed by other witnesses, mainly medical scientists, who asserted that medicine could do the job of assessment on its own. I was asked to testify at that hearing and favored the establishment of the study group, not because I thought that patients urgently needed protection but because I thought that medicine, as a profession, did. I suggested that only if we allowed ourselves to be exposed to public scrutiny could we expect to receive that level of public support which our enterprise, in my opinion, deserves.

In brief: I think it is a fatal strategy and tactic to assume that the public will support science, sight unseen, just because we think we are "the good guys," as indeed we are. When the cost of maintaining freedom is a question asked of the military and when the cost of proclaiming the

word of God is asked from the clergy, we would be utterly
naive to believe that the public will assent to our being
the only arbiters of what should happen in science and medicine regardless of cost. When the American public increasingly calls the Pentagon to account and the Catholic public
increasingly calls the Papacy to account, the time will
surely come when science shall be asked to account for itself too. To suggest, as Mr. Piel seems to do, that this
can be done through the peer review process is like suggesting that the public will be satisfied by having the
Joint Chiefs of Staff be the accounting body for Pentagon
activities or the College of Cardinals for Papal activities.
Those days are gone and we would be fools to plead for analogous privileges already denied those who claim the fiefdoms
of patriotism, freedom and eternal life in the hereafter.

How do these brief remarks, relate, concretely, to the
subject matter of this symposium? Let me attempt to construct a framework for our thoughts. No one questions the
right of individuals to contruct a theology or a philosophy
of life. No one questions the right of individuals, to construct a definition of freedom or patriotism. No one questions the right of individuals to construct a definition of
God. No one questions the right of individuals to assert
what they perceive to be facts or scientific truths.

The only issues in science, increasingly under public
scrutiny, are twofold:

1.   Given finite resources, how much should the public in-
      vest in an enterprise such as science?

2.   In how far (if at all) should any enterprises, in the
      name of freedom of inquiry, be allowed to infringe on
      the freedom of others?

To ask the first question, that of investments, is
simply to say that while there surely is a freedom to inquire there is nothing which says that it must be paid for
by others than the inquirer. That becomes an issue of priorities in achieving the common good. In such a decision
the scientists can surely plead his cause as being the
common cause, but, like the military, a slice of the pie
will be assigned which shall be determined by total needs of
the country and those needs are rightly determined at the
level of the public rather than by interest groups.

The second question is a question about the means by
which science is done. Obviously no one will hold that one
may investigate how far the average 20 year-old can throw the

javelin, by doing the experiment on a crowded beach. That
seems like a simple conclusion. Not so simple is the answer
to the question as to precisely what may be inquired into
that may affect others than the inquirers. That brings us
to such questions as the doing of research on those who can
not give consent, such as the mentally retarded. In general
it is held that subjects of inquiry must give consent if
they are to be inquired into and, if they cannot, then proxy
consent must be given by someone who has the wellbeing of
the research subject as much at heart as the subject would
have himself or herself. It is here that most of the public
debate arises. Should not the scientist, given the importance of the enterprise, be allowed to invade a subject's
privacy without consent? It is the old question of means
and ends. I know scientists who, on the grounds of the common good, would plead that some experiments should be done
even without consent because of the importance to all of the
information to be gained. Some of these same scientists
would violently object if the F.B.I. or C.I.A., claiming
the same common good, would plead the same privilege.

Hans Jonas has argued that the notion of freedom of
inquiry has come to us from pre-modern times and therefore
dates from times when inquiry involved only the inquirer.
Astronomy is given as the perfect example of doing science
which affects none other than the inquirer--although one
might still argue whether a telescope must be funded by the
public. I think there can be no doubt, however, that where,
<u>in the doing of the scientific inquiry</u>, the good (including
the privacy) of others may be affected, those others are
entitled to a say in whether the inquiry shall be done. It
simply will not do to make that determination by a so-called
peer group, as Piel suggests, for just as there may be a
peer group of the inquirers, so there can be peer groups of
those on whom inquiry is done.

It is sometimes said that science is now so complex
that only scientists can understand it and hence the value
of doing experiments to foster it. I would reject such a
notion with the simple comment that, if you cannot communicate to others what it is that you are doing, you probably
do not know what you are doing yourself. I think it is
this inability (or unwillingness) to communicate what we
are doing and why we are doing it which contributes to an
increasing questioning of our enterprise.

In brief then, there is no warranty for the misuse of
money or of people in scientific inquiry and the public,
rather than the scientific community, should be the judge
of that.

One final matter must be discussed. Are there some
areas of knowledge which should not be inquired into, because of the danger of knowledge itself? Robert Sinsheimer
seems to think so. He asks the question: what would be the
effect of finding extraterrestrial forms of life to which we
are as chimpanzees are to man? Those who believe in a God
and even in angels have surely lived with that burden, if
burden it be, for centuries, and I cannot see that it has
done them irrevocable harm. While I can agree with Sinsheimer that there can be knowledge which it is dangerous to
have (Shakespeare allowed as much) I must confess that any
knowledge can be dangerous. It is however the human condition to know and I cannot see the prohibiting of inquiry on
the grounds of knowing being dangerous.

More cogent I find Hans Jonas' concern when he claims
that the doing of science and the doing of technology are
now so inseparably intertwined that it is not possible to
separate them. There is then no such thing as pure knowledge which will be unused. I agree with his position but I
do not know what to conclude from it. It does not to me
constitute a warranty for prohibiting the acquisition of the
knowledge. It merely tells me that to such inquiry as does
not, in the inquiry, harm nature, but which may be dangerous
in its consequence, I would assign a very low priority. I
am therefore quite in favor of those who ask that science as
an enterprise provide possible impact statements for its
work. That inquiry which may have adverse impacts need not
be supported.

What is at bottom at stake is not the freedom to inquire
but the right to be supported in all inquiry. Let me then
summarize my own position: I think there should be an
absolute freedom to inquire if, in the process of the inquiry,
the freedom of others is not infringed upon. As to there
being a right to inquiry which others are obligated to
support, I think there is none such. It is at that point
that science must make a case for its activities and it is
that which we seem increasingly to be done badly. It is
very unhelpful, in that process of communicating with the
wider public on whose support we depend, to assert that a
right of inquiry is the same as its entitlement. Relating
what I have said to medical research I shall only add that
where the subject of inquiry is the human rather than
inanimate objects it is to be expected that all scientific
inquiry will be scrutinized even more closely. I hold
however that the principles I have outlined govern all
scientific inquiry.

2

Value Conflicts in Restricting Scientific Inquiry

Barry M. Casper

I'll begin by identifying two good reasons why our society and its institutions might wish to limit scientific inquiry:

(A) anticipated deleterious consequences of the inquiry itself; and (B) anticipated deleterious consequences of applications of knowledge obtained by the inquiry. Examples of the first include possible release of toxic substances to the biosphere as in the case of certain kinds of recombinant DNA research or chemical and biological weapons development or nuclear weapons testing or possible adverse psychological effects as in the case of certain kinds of experiments on human subjects. Examples of the second include possible misuse of tools that grow out of research as in the case of genetic engineering resulting from recombinant DNA research or proliferation of nuclear weapons resulting from a technique I'll discuss that would use lasers to separate the isotopes of uranium.

Value judgments are obviously crucial in the decisions that are made as to whether to proceed with such inquiries or to limit them. Each such inquiry has many potential consequences, some desirable, some undesirable, and at the time decisions have to be made about proceeding to invest large sums of money in a project, there is usually significant uncertainty about what its consequences will be. Weighing potential risks and benefits is like weighing apples and oranges--they are incommensurable. How they are to be weighted is a value judgment. Similarly, the policy implications of uncertainty--should one, for example, assume the worst case or some more probable outcome--involve value judgments.

What weight is applied to the apples and what to the oranges and what stance is taken on the implications of uncertainty is resolved in practice by who makes the

decisions and what their values are. And what often counts
is not individuals and personal values, but rather institutions and institutional values. What's good for General
Motors or Exxon or the Department of Energy or the House
Science and Technology Committee may not be good for Nader's
Raiders or the Sierra Club or the Arms Control and Disarma-
ment Agency or the Senate Foreign Relations Committee.

Few would disagree, in principle at least, with the
need to limit scientific inquiry when the inquiry itself has
potentially serious adverse consequences. Limitation of
certain kinds of research to specialized facilities designed
to prevent the release of toxic substances, restrictions on
the storage and transfer of toxic or radioactive materials,
and even outlawing certain kinds of research on human
subjects are accepted practices in the U.S. today.

To be sure, there is plenty of controversy when one gets
down to the specifics--when it comes to assessing risks and
benefits and weighing them to establish policy. This is well
illustrated in the current dispute over recombinant DNA
research. With significant disagreement over the potential
health and safety hazards, the key question has become who,
that is what institutions, should decide what policies to
adopt. There is clearly a tension between the community of
scientists, which has evinced a strong desire to police
itself, and other interests in our society--notably politicians. Each institution has its own policy biases. In the
case of DNA research the molecular biologists, proclaiming
the primary value of freedom of scientific inquiry, are
inclined to continue research on a relatively unrestricted
basis. The politicians, responsive to the health and
safety concerns of their constituents, don't want to be held
responsible if something crawls out of the lab before the
next election. They are more inclined to limit research or,
better yet, to pass the responsibility along to someone else.

Such limitations on research seem to follow a pattern.
There first develops a widespread public perception of an
immediate threat from the research. Frequently scientists
themselves contribute to arousing the public. In this way
an organized political constituency favoring limitations on
the research develops. For example, in the controversy over
nuclear weapons testing in the early 1960's, through the
efforts of Linus Pauling and other scientists, mothers began
to fear the effects of Sr^{90} in babies' milk. The value of
free scientific inquiry (unrestricted testing) was seen to
threaten other social values, such as personal health and
safety, in the minds of a politically potent constituency;
as a consequence, certain limitations on nuclear tests--

restricting them to underground--were eventually invoked.
Those institutions that favored unlimited testing--notably
the Atomic Energy Commission, the Joint Committee on Atomic
Energy, and the Joint Chiefs of Staff--were forced to modify
their position.

This reason for limiting research--its direct conse-
quences--raises many important and difficult issues' but
not the ones I wish to focus on today. Instead, I shall
concentrate on the possibility of limiting scientific
inquiry because of anticipated applications of knowledge that
might be obtained by the inquiry.

Here the threat is more distant and its dimensions less
clear. As a consequence, there is less likelihood that
political backing will develop in favor of limiting scientific inquiry on this account. Since the future has no
constituency, the politics of such limitations must be quite
different. If one is concerned with value conflicts in
limiting scientific inquiry, then a central question has to
be how intellectually astute and politically potent value
can be attached to the interests of future generations when
science and technology policies that will affect those
generations are made today.

In considering restrictions due to potential applications one is virtually forced to adopt a broader view of
scientific inquiry than just so-called basic research.
Limitations could come at any point in the path from
research and exploratory development to advanced development to deployment of a perfected technology.

To illustrate some of the issues that arise, consider a
specific emerging technology--laser enrichment of uranium.
Currently being developed in the United States at two government laboratories--Los Alamos and Livermore--and by one
private company--Exxon Corporation--this technique, if
successful, would provide a much cheaper, much more efficient means for producing reactor-grade uranium--the fuel
needed for nuclear power plants. It may also provide a new
and dangerous path to the proliferation of nuclear weapons
by making possible small clandestine facilities, whereby
nations that do not now possess nuclear weapons could
secretly produce bombs. In this way it may undermine international efforts to curb proliferation.

About a year ago I suggested that the U.S. declare a
moratorium on laser enrichment development and initiate
efforts to get other highly industrialized nations known to
be working on this technology to join us.[1] The responses

to this suggestion illustrate well, I believe, the kinds of difficulties a proposal to limit scientific inquiry on account of potentially deleterious applications will invariably encounter:

(A) "It's too early." We can't even be sure the technology will work at this point. If it does prove success-ful in producing reactor-grade uranium, it may not be suit-able for weapons-grade material. The Exxon people argue that their approach will very likely not be suitable for bombs, but the Livermore scientists, pursuing a very similar approach, say it will be. Before it is even known whether the technology will work, it's too early to be discussing its applications, say opponents of the moratorium proposal. A variant on this theme is the "let's cross that bridge when we come to it" argument. The time to discuss limitations intelligently will be when we understand just what the pro-blems are; then and only then can institutions appropriate to safeguard against these problems be devised.

(B) "We are not alone." The United States is not the only country with the scientific and technical personnel and resources to develop this technology. Several other nations, including the Soviet Union, France, and Israel, are known to have active laser enrichment programs. A unilateral cur-tailment of the U.S. effort will not by itself stop development of this technology.

(C) "We can't afford to live without it." There are two variants on this argument. The first is simply that the anticipated benefits of the particular application in ques-tion outweigh the risks or at least that they outweigh the likely benefits that would accrue from restricting its development. For example, if optimistic predictions prove valid, laser enrichment could significantly expand the supply of nuclear fuel for power plants, and thereby contri-bute to alleviating the anticipated crunch in the world's energy supply. A second variation on this theme is that closely related applications of the technology--laser chem-istry, for example, are bound to be pursued. Their tool is so powerful that it will surely have a multitude of benefits we cannot yet foresee.

These same arguments were made in the 1950's and 1960's debate over the testing of nuclear explosives and are heard today in the discussion about limiting recombinant DNA research. I'll respond to these arguments in the case of laser enrichment; these responses are generalized to other technologies as well.

I believe it's not too early to consider limiting or even halting the development of laser enrichment. Every new technology passes through a stage where go-no/go decisions are made. Roughly speaking this occurs after sufficient research and exploratory development have been completed that the potential of the technology can be realistically assessed, but before so much money has been invested that it acquires a momentum of its own. Given the nature of the institutions that make these decisions, in practice they are usually based on relatively narrow criteria of technical feasibility and economic payoff, as well as what might loosely be termed bureaucratic and special interest concerns. Rarely are societal implications, such as proliferation risk, a central consideration in such decisions.

Laser enrichment is now passing through that window in time when we know enough about its potential applications that we could use that knowledge to shape policy. Educated guesses can now be made, for example, about the potential of laser enrichment for clandestine production of weapons-grade material. But very difficult technological problems remain to be solved before a laser enrichment plant can be built. It is this stage between exploratory and advanced development that would seem to be the most propitious time to assess an emerging technology and integrate that assessment into decisions about its future course.

As for the argument that we are not alone, that is certainly true. Unilateral restraint by the U.S. is not enough. International agreements will be required to limit the development of laser enrichment and this is obviously true of proposals to limit other technologies as well.

Could we live without laser enrichment? Of course we could. It might cost Exxon several billion dollars in profits and we might have to live with more expensive and less efficient uranium enrichments, but alternative enrichment processes are certainly available. What is less clear is whether other applications of laser selection techniques, such as those of laser chemistry, could be pursued while laser enrichment of uranium was restricted. I'm not sure, but my guess is that uranium isotope applications will involve sufficiently specialized techniques that an effective separation could be made.

One way laser enrichment differs from the other examples of scientific inquiry I have mentioned--nuclear weapons testing and recombinant DNA research--is that objections to the latter two technologies involve both the health and safety implications of the inquiry itself and applications of knowledge the inquiry might produce, whereas no one is objecting to laser enrichment on health and safety grounds. The political opposition to nuclear testing and recombinant DNA research was/is based primarily on public perceptions of a threat to health and safety. The testing debate was resolved by the technical fix of moving the tests under- ground; this spoke to the safety concerns, but not to the applications of tests to ABM development or to the implica- tions of continued testing by the U.S. and the Soviet Union for the proliferation of nuclear weapons. Judging from the current status of congressional legislation, I suspect a similar technical fix will occur in the recombinant DNA case, allaying public fears about safety, but not speaking to the potential misuse of genetic engineering. As for laser enrichment, there is not likely to be any public controversy at all; in the absence of any direct threat to the public from the research, no politically meaningful opposition will arise and those institutions developing this technology will be free to proceed.

There has been much talk of "technology assessment," and "early warning," but so far the mechanisms we have created such as Congress' Office of Technology Assessment or environmental and arms control impact statements haven't much affected the politics of technology--haven't forced the traditional decision-making institutions to modify their practices in the R & D stage of new technologies, which I think is the critical time. For example, in the case of laser enrichment, even the minimal first steps of an arms control impact statement or a detailed technology assessment by OTA have not been prepared. Even if they were, I am dubious that they would have much effect.

Of the two good reasons I cited at the beginning of my remarks why society might wish to limit scientific inquiry, only the first, direct consequences of the inquiry itself, has much relevance today. I think that's a serious defi- ciency in our present institutions and in the values that ought to be taken into account when decisions are made about our technological future.

References

1. B. M. Casper, "Laser Enrichment: A New Path to Prolifer- ation?" Bulletin of Atomic Scientist, 33 (1) (1977): 28-41.

3

Panel Discussion

Should Scientists Limit Their
Own Research?

Casper: I would like to clarify one thing I said: we don't
have good mechanisms for taking into account the potentially
deleterious applications of knowledge that we obtain from
scientific inquiry. I did not mean to imply that an effective response is simply for scientists to limit their own
research. My own feeling is that that by itself is not an
effective answer because the "we are not alone" argument pertains to not only U.S. science vs. other countries' (that is,
if we stop doing something, scientists in other countries
will do it); it pertains to the scientific community in the
U.S. as well. There are very few developments which would
be affected if any particular individual scientist were to
refrain from participating. Scientists are not alone; they
are replaceable. There are other scientists and engineers
who will do the job if they don't. So I don't think that
that's the answer.

   The question of what new mechanisms or what new institutions ought to be developed to introduce consideration of
potentially deleterious consequences of applications of knowledge into the decision-making process is a complicated one.
I don't think our present mechanisms and institutions work
very well. The kinds of activities that are termed "whistle
blowing" are part of the answer. That is, scientists who
are concerned about applications of knowledge can refrain
from doing it, but a much more useful thing would be to alert
the public to what they see about these applications, and to
try to develop political constituencies. That's not easy
though; the public tends not to get very excited about

These remarks are from the discussion which followed the presentation of Hellegers, Casper, & Back at the AAAS symposium.

potential risks except ones that are very immediate threats, like something crawling out of the laboratory. For example, it is very difficult, but not impossible, to interest the public in the proliferation risk of laser enrichment or in the potential genetic engineering applications of DNA research, as opposed to its immediate health and safety effects. Scientists can play an important role in that, but that's another, long story.

Hellegers: I trust that you would go extremely slow in saying that you may not inquire into something because the knowledge acquired might be dangerous. I think that is open to a tremendous amount of misuse. I would agree with Dr. Casper that the scientist ought to state publicly and inform the public what possible deleterious effects might occur. But the notion that one would prohibit the acquisition of knowledge, because the knowledge might be dangerous, is repugnant to me, although I agree that you don't need to fund it. I don't think there is a right to funding, and if you believe that, in fact, an activity might be a dangerous enterprise, you could set a very low priority. I'm more interested in things that are not dangerous than things that are dangerous. But the notion of prohibition and the notion of nonfunding, I think, are quite different notions.

Casper: But the really interesting question is the political and economic question of who it is that decides priorities. In our society, as the example of enrichment suggests, those decisions are usually made by relatively insulated groups. In the case of laser enrichment, Exxon Corporation, the Department of Energy, the Joint Atomic Energy Committee, and the House of Science Technology Committee are promoting and paying for development of the technology.

Hellegers: And broader, though, than the peer system.

Casper: Okay, but you say that the way the process ought to work is that low priority ought to be given to these activities which may have adverse consequences. I'm saying those institutions in our society which make the decisions about what priority to give to those various research and development activities tend to do that on the basis of factors such as: will it work? will it make a profit? will our laboratory continue to exist and prosper?--criteria like that--and tend not to take much into account the potentially adverse consequences. I think that that's just a political fact of our institutions. I would say that at the present point in the development of laser enrichment we know enough, although we don't know for sure whether the technology is going to work. What we know ought to influence our policies. For

example, we know enough, I think, that we should not be supporting Livermore Laboratory to develop weapons grade applications of this technology. We're not doing that. In fact,
for a bureacratic institutional reason the Livermore Laboratory is not only developing laser enrichment for nuclear fuel,
but it's going ahead and trying to develop laser enrichment
for highly enriched uranium that could be used for bombs.

Hellegers: But it seems to me there is a difference in
determining how knowledge is gained and how it shall be used
technologically and that there is great danger in that transition. That's why I agree with Hans Jonas on that, but I
don't find a reason to prevent acquisition of knowledge.[1]
I think that's where we may differ.

Casper: I think that when you start considering the question
of limiting scientific inquiry generally because of possible
adverse applications of the knowledge, you can't consider
basic research or acquisition of knowledge separately from
the whole development process. You've got to think of it as
a continuum; you've got to think of where in that continuum
from research to applied research to exploratory development
to advanced development to deployment you might consider
limitations. The tendency in our society is to wait until
technology is deployed and then try to regulate it, as we
say. I think that is a mistake. Once it gets to the regulation stage, the deployment stage, technology is out of hand.
What I was trying to suggest in my paper is that I wouldn't
go way back to the early stages of research where you really
can't anticipate very well what these consequences are. I
think the most salient, the most important, point to begin
considering limitations is in that stage between exploratory
development and advanced development where the money really
begins to be put in, and that's not usually the way it
happens. I think we are at that stage in recombinant DNA
research. I'm not saying that we should stop recombinant
DNA research, but I think now's the time to start considering
in a more subtle, more sophisticated way just what kinds of
research might be permitted and what kinds of research we
might want to restrict because of what we can anticipate.

Hellegers: On the basis of the knowledge or on the basis of
the harm done in the inquiring?

Casper: On the basis of what we anticipate to be the applications of the knowledge.

Hellegers: So that the knowledge might be dangerous.

Casper: Right.

Hellegers: Oh, I would be strongly opposed to that; that, I think, is where our dividing line is. I think I'm heartily in favor of the regulation of technology in terms of where it does harm. But I think it would be terribly dangerous to say that I will cease investigation or inquiry because something harmful might come from that knowledge. Because at that point you anticipate humanity.

How Should Technological
Assessment be Done?

Casper: Using DNA as an example, there are many institutions that could do such assessments. We have one in Congress now, the Office of Technology Assessment, which would seem a natural. The important thing is to develop a public record of what the potential applications of recombinant DNA research are and lay that on the line. There are a number of promising techniques that have been proposed to do that. For example, a mediation proposal that Nancy Abrams and Steve Berry put forth would get proponents and opponents of the DNA research together, have them explain each other's positions to each other to the satisfaction of one another, and then write a document which says, here is what we agree on; here's where we disagree; this document would include a discussion of the potential applications and their beneficial and adverse implications.[2] Such a public assessment could be required by legislation. It's important to involve a rather large range of interest groups--in this case the scientists in the molecular biology community who first brought this issue to light but also many others--to produce a document which lays out what are the potential implications of the technology. After such an assessment is produced, however, it's not at all obvious how it will be linked to the policy development. I personally believe the links have got to come from the development of political constituencies. Having the information is not enough. It is an important first step, but it is going to take interest groups to relay this information to the public. It is going to take new uses of the media to relay this information to the public in order to develop the kind of public concern that our political process listens to. That's not easy because, as I asserted in my talk, the fact that we're getting just safety legislation out of the Congress--in dealing with DNA research --that is, a bill that deals only with the safety hazards--is probably the result of an accurate reading of public sentiment by the Congress. That's what most people are really concerned about. It's only a relatively small number of people like yourself who are concerned about what's going to happen 10, 15, 20 years from now with genetic engineering. The really difficult problem is to make this information

about the potential threat of, say, genetic engineering
available to the public in a form that they can understand
and react to. The answer is development of political
constituencies, but I don't think it's easy to do.

Back: I think the problem is a little broader, because it
really concerns the development of science. There are a lot
of issues in science. But the position of scientific
research is so weak in society that we can very easily throw
out the baby with the bath and kill all of scientific
research at the same time. Take another example of some
other technology. We always discuss the kinds of technology
we don't like. But take, for instance, an improvement in
abortion techniques which makes it easy for any woman to perform an abortion herself--something which many people feel
is murder. I personally don't, but some people do, and are
very much opposed to it. In private life this might be
equivalent to the potential of the small nations having an
atomic bomb. Every woman could then have an abortion, but
there are ethical interests involved; some people would be
against this scientific advance very much. Now should we,
therefore, until we have seen all the implications involved
in abortion, for the future of society, for the future of
democracy, have a moratorium on all research on abortion;
should we have a commission, with some members appointed by
the pope, some appointed by various research laboratories,
some appointed by other interested people, who could come to
an agreement, who would sign a statement which would be presented to Congress? You know how Congress would vote on
that: after that time we would have no more abortion
research, we would have a complete moratorium. We have one
such issue after another, and with one such issue you can
destroy scientific progress very easily. Now this may be a
good thing, but I don't think AAAS should have a position on
this issue.

Hellegers: I may be the only one here who has served on a
commission for both a pope and president. I wouldn't go
that route at all if I were you, that would be the last thing
I would do. I think that Dr. Casper is fundamentally correct
that it should go through the political process and political
lobby. That is what the Kennedy Institute in Georgetown is
in part about; that is what the Hastings Center certainly
does in cooperation with people. I still think though that
even if we do this in institutions, and I would wish that
there were many more set up of that kind that did technology
assessment, you would still have a problem of taking that
one step further, namely to the point where the public
insists that something be done for good or ill. I think the
public should suggest that NSF or NIH--or a similar agency--

be made to spend 1% of its funds on this kind of work. What
I mean is simply that an automatic rider be attached to fund
some reflection on the consequences of the work funded. For
example, for every 99 bucks spent on conquering cancer, $1
must be spent reflecting on the consequences of conquering
cancer (those consequences are strokes, you know, because if
you've got to go, you'be got to go somehow. Even with
obviously good work like death prevention you buy, eventually, senescence.) Nobody does that kind of technology
assessment right now. That kind of technological assessment
of the consequences of scientific progress should be encouraged. I agree with Dr. Casper and ultimately you can only do
it through the public and the political sector. And I would
hate to see it be done in any other way. I think that if
the Congress right now thinks only in terms of safety in the
recombinanct DNA kind of situation, I think you are right
that it perceives that's where its lobby is, but I think it
might also be because they, in fact, have no intention at
this juncture to interfere with the acquisition of knowledge,
regardless of potentials for genetic engineering and so
forth, and that it would say fine, let's acquire the knowledge first and then regulate how we can use it. And I
think what all of us have to carefully think through, is at
which point do you talk about regulation and interference.
And I have just simply made the plea that it not be done on
the basis of the fact that knowledge is dangerous. Because
if we once go that route I think one might very well . . .
you know.

Casper: I'd like to make one final comment which relates the
question about how assessment might be incorporated into legislation. There is an idea that was around a year ago--it is
pretty near dead now--called the Science Court, which has
what I consider to be all sorts of fatal deficiencies. But
one good feature of it is that it would involve a range of
groups with different opinions about what policies on a particular issue ought to be and give them funding. That is, it
would bring much more into balance the kind of imbalance
that was cited between the funding that opponents and proponents of recombinant DNA research have. I would hope that
the kind of assessment that would be incorporated in legislation would bring into the process of assessment representatives of groups with a variety of views about what policy
ought to be, and give them sufficient funding to allow them
each to make their case effectively.

Where Does the Leadership Come From?

Casper: The kind of leadership and decision-making processes
we have now are quite insulated from the public. The people

that are making the decisions about whether we should go
ahead with laser enrichment or whether we should go ahead
with recombinant DNA research (and I don't mean to imply that
the decision is necessarily between a total prohibition and
full speed ahead--in both instances there are options in
between) are largely insulated from the public. There are a
variety of alternatives. You may feel that if such decisions
were left to popular referenda, it would tend to slow things
down very much. I think you are probably right about that.
But it depends a lot on how things are done, what kinds of
information, in what form, are provided to the public about
what the policy options are and what the implications of
various options are. I don't think we do a good job at all
--the media in particular don't do a good job--in informing
the public. But who has the money to pay for more media
coverage? Who has the money to bring the more diverse set of
views into the process in a way that they can actually influence the process? I think that's the biggest problem.

Back: Well this organization was founded to provide the
kinds of leadership you are asking for. Maybe we should
have mechanisms within this organization to provide it and
not to attack science.

John Edsall: I might say with regard to what Dr. Casper was
just saying that I think that the isolation of the leadership
from the public is much more far reaching in the area of
weapons development than it is in most other areas. I do not
think that the leadership in the case of recombinant DNA
research has been insulated so widely from the public. I
think there has been a great deal of public interaction there
and a great deal of publicity about facts and about what the
issues are. I think there is a very strong sentiment among
the scientific community generally that any real violation
of the guidelines for recombinant DNA at the present time
would be condemned by the scientific community at large, and
this is a powerful inhibition which is perhaps just as
effective as legislation, maybe, for the time being.

Unindentified: I think we should concern ourselves not with
the acquisition or the limitation of the acquisition of
knowledge but in the application of that knowledge.

Hellegers: I think that you are expressing pretty much my
position--the knowledge itself and the acquisition of the
knowledge I have nothing against--but you have not yet

settled by the way you said that what kind of acquisition of knowledge you are going to fund in preference to what kind of other acquisition. That's where the action is at, because the individual scientist staring at his bellybutton and so forth is no longer where it is at. So it is, I think, the priorities in the financing that really govern this.

Has the Priority System of Science Resulted in Neglecting the Practical Problem of the Consequences of Scientific Research?

Hellegers: Let me try and answer. I think the problem actually is what we are dealing with in the latter part of this discussion, and when we talk priorities, we are dealing with values, right? Now for a very long time I think it has been a tradition to get rid of values, you know the old notion of my values are as good as your values, my ethics are as good as your ethics--all that kind of thing. Value education and systems by which values were determined came out of the departments of philosophy, departments of ethics and so forth, and if you believe that science is hurting, boy, you ought to be a philosopher. They are really in bad shape, and they are really in bad shape because science has in rather a systematic way denied their usefulness. And now science itself is beginning to say gee, they were more useful than I thought they were, and I wish I could do some of that kind of work. I happen to know this because I've worked in Edinburgh, in Paris, and in Holland, at John's Hopkins, at Harvard, Yale and now at Georgetown, and it is to me remarkable how different those experiences are. And the funniest of the lot is Georgetown because you are suddenly in the middle of a bunch of Jesuits who say that's what we've been telling you for centuries. So I think one of the things that one has to do as a lobby is to take a much less harsh stance against the humanities than we have done, and it is fascinating that in the Federal Government right now if anybody is concerned with the kind of things we are talking about here it is the National Endowment for the Humanities. NIH hasn't gotten in there yet. It's NEH that has been supporting most of this kind of work, and the federal government doesn't know quite where to place it. There is enabling legislation for the National Science Foundation to do technology assessment and bio-ethics. NEH has such language. The Congress put it in. When you go to those constituents the tendency of NEH is to say, yes, but you know we are financially responsible for Shakespeare, not for clearing up the mess of medicine--why doesn't medicine do that. When you go to NIH, NIH says, yes, but you know we are responsible for precisely doing this biology and not for assessing its values. The net result becomes that what we are talking about here, which is the ordering of priorities in things which affect both science and human values, falls between all the cracks, right? And that is where I see the technical problem right now. That can only be changed by the kind of enabling legislation that can be done by any congressman or any senator that would say to an NIH or where ever you wish to place it bureaucratically, you are now charged with doing it. And that can be done.

Casper: I guess I'm skeptical that the answer to the kinds of problems that we are discussing here today has much to do with additional funding either to social science as you suggested or to philosophers. I think it's got more to do, at least as far as money is involved, with somehow resolving the disbalance between the resources of various elements of our society and giving a more diverse set of policy options the kind of public airing that they deserve.

Should Persons Who Violate Ethics in Regards to Human Experimentation be Professionally or Legally Prosecuted?

Hellegers: Boy that's a biggy. Let's start off by saying that there isn't even agreement right now on what is ethical and unethical research. So quite apart from saying, you know, disbar people, stick them in jail and so forth, much more has to be done just simply on the application of existing ethical systems to biology at all. Even that hasn't been thought through. Now temporarily what has happened due to political process is that the National Commission for the Protection of Human Subjects was set up. That obviously is not going to be able to come up with a normative ethical set of standards for the simple reason that when you create a commission, you take them coming out of different ethical systems to start with, then you throw them together into one commission. What you come up with is obviously a compromise between loads of ethical systems that factually do not consist of one whole. So what happens then is you write regulations. Input into the regulations is quite clearly allowed. You know, a lot of people who complain about this never bother to read the federal regulations and never bother to put their inputs in. And most of those regulations, if you read them, have been regulations suggesting an enlargement of the body of people who shall look at the research before it is done. Institutional review board type things. Would there be harm to have an ethicist on such a review board? Absolutely not. I think it would be very advantageous. But I would not like to see just bioethicists

sitting on that. Because then the question would be what
are they in bio-ethics. Are they utilitarian, are they
John Rawls' distributive justice types. Right now one of
our major concerns at Georgetown is to study the impact of
institutional review boards and their composition on the
projects they review. The institutional review board can
through its predetermined ethic set up the whole thing to
come out ethical according to the ethical system they follow,
which can vary anywhere from an arch dontologist to arch
teleologist. So yes, I would have a bioethicist on the commission. Two would be fine, great, depending on how large
the group is. I think that you also need some unbrilliant
people--what used to be called in England twelve good men
and true. I would rather trust the good and the true than
either the Ph.D. in science or the M.D. or the Ph.D. in
bioethics.

Can Peer Review be a
Democratic Process?

Hellegers: That is where the trouble started. I think that
when it was determined that the peer review system would set
the priority for funding projects on the basis of their
scientific merit, the whole connection of the social value
and possible social consequences of research with the
possibility for funding was lost. So if you say that the
study sections of NIH shall rank projects on the basis of
scientific merit, you lose the ability to do what you want
to do, that is, to have some idea of whether the highest
rated scientific projects are the most useful for mankind.
With the present rating system that is not necessarily so.

Who Controls Public Knowledge?

Unidentified: The whole question of regulation of scientific
research really goes back to the question of ownership. We
are in an institutional framework now, wherein we have a
sponsor, the government, that funds a great deal of research,
that funds education that trains the scientists who then go
on either to assume higher public institutional jobs or jobs
in private industry. A lot of the knowledge being acted
upon by Exxon and others is knowledge that emerged out of
the public treasury. Who then has the responsibility to
control that knowledge? Does it belong to the people who
provided the basis for it in the first place, or do we
continue to use the framework of the solitary thinker and
say this knowledge has burst into the air and whoever
catches has the right to use it.

Hellegers: Again I think it is a question that is political.
For you see when you make the blithe assumption it is the

government that has trained these people that Exxon now uses, I would turn around and say to you, who is the government? The public fundamentally has the problem of regulating the public. Now the question is who should be our intermediaries. Shall it be the judicial branch, the executive, or where the dickens does it lie? I still think that comes back then to say, when all the questions are asked that have been asked here, you have met the enemy and it is us. We have to do the lobbying, and if it costs funds to lobby, some people march across the country. They get very effective reportage on the way from the press. I simply cannot see how it can, at least in the United States, be other than to say we shall do the regulating through our elected representatives. That is the only regulating body in existence; whether it does or does not regulate depends on us.

Casper: In many cases today it does not. In the case of many technologies, decision-making forums are very much insulated from the public.

Hellegers: Then the public has to go out and do something.

Casper: The question is what kind of system do we have in the U.S. today. I would say one in which technological policy is to a large extent insulated from the public. We have to change that.

Hellegers: That is right. And the public is to blame. It richly deserves what it is getting. One gets the government one deserves.

4

Freedom of Scientific Inquiry and the Public Interest

Hans Jonas

Freedom of inquiry and the idea of it are precious to the Western world as part of its general regard for freedom. Freedom of inquiry is claimed, granted, and cherished as unqualified on the premise that inquiry as such raises no moral problems. Let us take a look at this all-important premise, bearing in mind that "inquiry" today means preeminently <u>scientific</u> inquiry in the technical sense.

What are the points of contact between science and morals? At first glance there seem to be none, beyond the internal morality of keeping faith with the standards of science itself. Its sole value is truth, its sole aim the knowledge of truth, its sole business the pursuit of knowledge. This, to be sure, imposes its own code of conduct which can be called the territorial morals of the scientific realm: abiding by the rules of evidence and method, not cheating oneself and others, for example, by sloppy reasoning or experiment, let alone falsifying the latter's outcome—in short, being rigorous and intellectually honest. Ethically this amounts to no more than the command to be a good rather than a bad scientist (that is, when a scientist, be a scientist!) and implies no extrascientific commitment. The same is true for the personal virtues of dedication, persistence, discipline, and the strength to resist one's own prejudices—again simply conditions of success within the vocation, if also praiseworthy qualities in general. Finally, the duty of sharing one's results and evidence with the scientific community seems to lend a social and public dimension to intrascientific morals; but in fact, given the increasingly collective nature of the scientific enterprise, intercommunication belongs—even for the individual investigator—to the technical conditions of doing science well: it still leaves the scientific morality strictly "territorial" and as yet stipulates no obligation of the scientific fraternity beyond itself.

Reprinted from <u>Hastings Center Report</u>, 6 (1976): 15-17

We feel, of course, that this cannot be the whole truth. It may have been true as long as the contemplative sphere and the active sphere were cleanly separate (as they were in premodern times), and pure theory did not intervene in the practical affairs of men. Knowledge could then be considered a private good to the knower, which—being merely a state of mind—could do no harm to the good of others, as it sought only to comprehend but not to change the state of things. Its dissemination, indeed, was sometimes regarded by public powers (such as the Church, but sometimes also the state) as dangerous to the good of the many, for example, by undermining their faith; but a quasi-automatic protection against this lay already in the esoteric character of higher learning as such, which confined its reception to the few; and those few had mainly to defend the right to their own thought against custodial claims on their souls, seeing that it did not trespass on things in the outer world. And, after all, even broadcast widely among the untutored, ideas have at most persuasive and not coercive force.

Merging Theory and Practice in Modern Science

All this lapsed with the rise of natural science at the beginning of the modern age, which entirely altered the traditional relation of theory and practice, making them merge ever more intimately. Even so, the fiction of "pure theory" and its essential "innocence" persisted: under the banner of the general freedom of thought and speech, as distinct from deeds, scientific inquiry too claimed untrammeled freedom for itself on the same distinction—in curious concurrence with the promise of eventual usefulness, which contradicts the plea of theoretical insularity. It took the Industrial Revolution to fulfill the promise of usefulness on a large scale. Until then, the social charter of science still rested on the uninherited dignity of "knowledge for its own sake," now joined with the principle of toleration for all thought and belief (including the right to err). So deeply is this two-fold respect ingrained in the modern mind that even in today's vastly changed situation, few things sound more odious to Western ears than "interference with the freedom of inquiry."

Sincere as this homage to disinterested knowledge may often be, it would be hypocritical to deny that in fact the emphasis in the case for science has heavily shifted to its practical benefits. From sometime in the nineteenth century onward, and accelerating in ours, there was an increasingly irresistible spill-over from theory, however pure, into the vulgar field of practice in the shape of scientific technology. Belatedly and almost suddenly, Francis Bacon's (1561-1626) precocious directive to science to aim at power over nature for the sake of raising man's material estate had become working truth beyond all expectations.

Though the "esotericism" of the proliferating branches of knowledge has even heightened and keeps heightening to the point of virtual inaccessibility to all but the insiders on each twig, yet the public impact of their recondite cerebrations is enormous: an impact not, as it was at most before, on opinions but on conditions and ways of life. And therewith the subject of "science and morals" begins in earnest. For whatever of human doing impinges on the real world and thus on the welfare of others is subject to moral assessment. As soon as there is power and its use, morality is involved. The very praise of the benefits of science exposes science to the question of whether all of its works are beneficial. It is then no longer a question of good or bad science, but of good or ill effects of science (and only "good science" can be effectual at all). Is it responsible for either? Clearly, taking credit for the benefits means also taking blame for the damages; it would be better for science to do neither but this option may be closed. Apportioning praise and blame can be an idle exercise, but it is not when a social privilege—and the freedom of inquiry is nothing else—is implicated in it. Thus it is not idle to ask: if technology, the offspring, has its dark sides, is science, the progenitor, to blame?

The simplistic answer is that the scientist, having no control over the application of his theoretical findings, is not responsible for their misuse. His product is knowledge and nothing else: its use-potential is there for others to take or leave, to exploit for good or for evil, for serious or for frivolous ends. Science by itself is innocent and somehow beyond good and evil. Plausible, but too easy: witness the soul-searching of atomic scientists after Hiroshima. We must take a closer look at the interlocking of theory and practice in the actual way science is nowadays "done" and essentially must be done. We shall then see that not only have the boundaries between theory and practice become blurred, but that the two are now fused in the very heart of science itself, so that the ancient alibi of pure theory and with it the moral immunity it provided no longer hold.

Denying the Moral Immunity of Science

The first patent observation is that no branch of science remains whose discoveries are devoid of some technological applicability. The only exception I can think of is

cosmology: the expanding universe, the evolution of galax-
ies, big bang and black holes—these are matters for knowing
only and for no possible doing on our part. It is worth our
reflection, and surely no accident, that the first science
of all, astronomy, the contemplation of the heavens, is also
the last to be purely science. Every other unraveling of nature by science now invites some translation of itself into
some technological possibility or other, often even starts
off a whole technology not conceived of before. If this were
all, the theoretician might still argue his sanctuary this
side of the step into action: "That threshold is crossed
after my work is done and, as far as I am concerned, could as
well be left uncrossed." But we must remind him that he
could not have done the first, "pure" part without massive
arrangements from outside under whose broader roof his role
becomes part of a contractual division of labor. What is the
true relationship?

     First, much of science now lives on the intellectual
feedback from precisely its technological application. Second, it receives its assignments from there: in what direction to search, what problems to solve. Third, for solving
them, and generally for its own advance, it uses advanced
technology itself: its physical tools become ever more de-
manding. In this sense, even purest science has now a stake
in technology, as technology has in science. Fourth, the
cost of those physical tools and their staffing must be un-
derwritten from outside: the mere economics of the case
calls on the public purse or other sponsorship; and this, in
funding the scientist's project (even with "no strings attached") naturally does so in the expectation of some future
return in the practical sphere. Here there is mutual understanding. With nothing shamefaced about it, the anticipated
payoff is put forward as the recommending rationale in seeking grants or is specified outright as the purpose in offering them. In sum, it has come to be that the tasks of science are increasingly defined by extraneous interests rather
than its own internal logic or the free curiosity of the investigator. This is not to disparage those extraneous interests nor the fact that science has become their servant, that
is, part of the social enterprise. But it is to say that the
acceptance of this functional role (without which there would
be no science of the advanced type we have, but also not the
type of society living by its fruits) has destroyed the alibi
of pure, disinterested theory and put science squarely in the
realm of social action where every agent is accountable for
his deeds. Add to this the pervasive experience that the
pragmatic implications of scientific discoveries prove irresistible to the marketplace—that what they show can be done,
will be done, with or without a prior compact—and it is

abundantly clear that no insularity of the theoretical realm
still saves the scientist from being the generator of enormous consequences. While technically it is still true that
one can be a good scientist without being a good person, it
is no longer true that being a good person begins for him
outside his professional work: the very doing of it entails
moral questions already inside the sacred precinct.

How much "inside" becomes clear when we reflect on the
third point in our list, the employment of physical tools--
that is, on how the scientist gets his knowledge. It is
then borne in on us that doing science already includes
physical action; that thinking and doing interpenetrate in
the very procedures of inquiry, and thus the division of
"theory and practice" breaks down within theory itself.
This has an important bearing on the hallowed "freedom of
inquiry," after inquiry has become essentially "research."
There was a time when the seekers after knowledge did not
need to dirty their hands; of this noble breed the mathema-
tician is the sole survivor. Modern natural science arose
with the decision to wrest knowledge from nature by actively
operating on it, that is, by intervening in the objects of
knowledge. The name for this intervention is "experiment,"
vital to all modern science. Observation here involves
manipulation. But the granting of freedom to thought and
speech, from which that of inquiry derives, does not cover
action, even if subsidiary to thought. Action is always
subject to legal and moral restraints. However, two properties of classic experimentation still ensured the "innocence" of this kind of action internal to scientific
inquiry: that it dealt with inanimate matter, and on a
small scale. Not real thunderstorms but discharges from
condensers are generated to learn about lighning. Simu-
lating models, containable within the laboratory, substitute
for the real thing. In that respect, the insulation of the
cognitive arena from the real world still holds.

Both these guarantees of harmlessness and therefore of
freedom in experimentation have lapsed with certain more
recent developments in science. As to scale, an atomic
explosion, be it merely done for the sake of theory, affects
the whole atmosphere and possibly many lives now or later.
The world itself has become the laboratory. One finds out
by doing in earnest what, having found out, one might wish
not to have done. And as to experimentation on animate
objects, which came with the younger biological sciences, no
one surrogate will do, no vicarious model, but the original
itself must serve, and ethical neutrality ceases at the
latest when it comes to human subjects. What is done to them
is a real deed, for whose morality the interest of knowledge

is no blanket warrant. In either case of experimentation,
that of excessive magnitude and that performed on persons,
to which others could be added, the protective line between
vicarious and genuine action is obliterated in the execution
of research itself--rendering somehow obsolete the conventional distinction of "pure" and "applied" science. Not only
the "what," already the "how" of cognition straddles the
line. Application takes place in the inquiry itself.
(Where it does not directly, as in theoretical physics, the
experimental base is vitally drawn upon.) It follows that
the "freedom of inquiry" cannot be unqualified.

Qualifying Freedom of Inquiry

We are with reason touchy about interference with this
freedom, not only because it had once painfully to be
wrested from earlier thought control and is thus a precious
and vulnerable possession, but also because we have before
our eyes its shameful repression in the East. Yet we must
remember that the high privilege of theory had its own
theoretical foundation in the distinction of thought and
action and is really conditional on it. Never has absolute
freedom been claimed for action, and surely never has it
been accorded to action. Thus to the extent that the operation of science becomes shot through with action, it comes
under the same rule of law, social censorship, moral
approval or disapproval, to which any outward acting is
exposed in civil society. And, of course, its own internal
morals cease to be purely "territorial": the very means of
getting to know can raise moral questions before the
"extraterritorial" question of how to use the knowledge so
obtained poses itself.

It would weaken the case to illustrate it only with
notorious atrocities. It is easy to elicit unanimity on
such examples: that one must not, in order to find out how
people behave under torture (which may be of interest to a
theory of man) try out torture on a subject; or kill in
order to determine the limit of tolerance to a poison, and
the like. We are referring, of course, to the deeds of
physicians (some of them prominent ones!) in Nazi concentra-
tion camps. Here was "freedom" of inquiry as shameful as
its worst suppression. But we know too well, or believe we
know, that the perpetrators of such scientific experiments
(yes, scientific they could have been) were despicable and
their motives base, and we can wash our hands of them. We
may go further and question whether in these cases the knowledge sought after is a legitimate scientific aim in the
first place; and if we conclude that it is not (for which
there would be good reasons) then we could say that we are

not really dealing with a case of science, but with one of
human depravity. But our problem is not with crooked or
perverted science but with bona fide, regular science.
Keeping, then, to indubitably legitimate and even praiseworthy goals, we ask whether in their pursuit one may, for
example, inject cancer cells into noncancerous subjects, or
(for control purposes) withhold treatment from syphilitic
patients--both actual occurrences in this country and both
presumably helpful to a desirable end. I do not rush into
an answer, but I do say that here moral and legal issues
arise in the inner workings of science, long before the
question of application arises--issues that crash through
the territorial barriers of science and present themselves
before the general court of ethics and law. To the public
authority of that court even the vaunted freedom of
inquiry must bow.

5

Scientific Research
Determining the Limits

Gerard Piel

A portrait of Benjamin Franklin, steel-engraved in Paris
during the American revolution, bears a legend in Latin which
translates: "He wrested the lightning from the sky and the
scepter from the tyrants!"* As a natural philosopher, Franklin drew none of our latter-day distinctions between science
and engineering, and he recognized a direct connection between the natural and the social sciences. He had founded
the American Philosophical Society long before, in 1743, to
be "Held at Philadelphia for Promoting Useful Knowledge."
Franklin knew that the promoting of such knowledge—today we
call it scientific—is a subversive activity. This is the
knowledge that sets men free.

Franklin would know how to dispose of the proposition
that it is somehow desirable, necessary and possible to set
limits on the work of science. He would dismiss it with one
of his proverbs that would at once illuminate the dark recess
in the human spirit from which such propositions come and set
the liberated spirit to laughing at its meanness and fear. In
his absence, we must take the long way around to an understanding of what was to Franklin self-evident.

The case for limits is argued today principally from the
hazards that attend certain kinds of scientific work. There
is the danger from experiments in molecular genetics, for
example, that by accident or by malevolence an irremediable
virus might be set loose among the people. I think the anxiety that charges this argument from practical contingencies
has its roots in deeper moral questions implicit in this
work. It is one thing for our imperfect societies to find a
dilemma in the choice of life or death presented by the pres-

---

*Eripuit caelo fulmen; sceptrumque tyrannis!

ence of nuclear weapons in the armories of national states. The dilemma presented by recombinant DNA arises from the work itself. Assuming that man will not choose the lunatic option of suicide by thermonuclear war, should he presume to seize the control of Evolution that is now so nearly in his hands?

Against such misgivings, I shall argue that no limit can or should be set upon scientific inquiry. I begin by urging you to consider the grounds of your liberty as self-governing citizens.

"We the People" ordained the Constitution. Not as an afterthought but in reaffirmation of our sovereignty, the First Amendment enjoins the Congress to "make no laws" abridging the freedom of speech. The authors understood that freedom to be absolute: not compromised by the claims of any competing interest, not even national security.

In our time, Alexander Meiklejohn has had to remind us that freedom of speech is the essence of self-government. It was Meiklejohn who first discerned the paradoxical duality of our citizenship. We are at once the governors and the governed. When, as governors, we take up the burdens of the public business our sovereign right to speak and, equally important, to hear is sanctioned by the unambiguous language of the First Amendment. We spend most of our lives, however, as people governed, in the pursuit of our private interests and ambitions. In this the lesser of our two roles as citizens we are quite properly subject to the laws we elect the Congress to write. Our freedom in this role is protected by the Fifth Amendment. In accordance with the terms of that amendment, we can be deprived of our freedom, including the freedom of speech in our private interest, by due process of law.

At this point, I have to concede that this view of our freedom is not the law of the land. The First Amendment did not get before the U.S. Supreme Court for interpretation until our country had joined the anarchy of nations as a full participant in the first quarter of this century. In the Schenck case, involving spoken and written opposition to the World War I draft act, Justice Holmes propounded for a unanimous court the "clear and present danger" doctrine that placed the First Amendment on one pan of the scale of justice to be balanced against the Government's duty to provide for the national defense on the other. The Yankee from Olympus is more famous for his dissent in the Gitlow case, in which he argued the same doctrine the other way. He had done his worst, however, and his misunderstanding of the nature of our citizenship is not only the law of the land but is cele-

brated as the popular ideal. Holmes' doughty eloquence on
the freedom of the speaker to hawk his thought in the marketplace of ideas confounds the public freedom, which is required for the purposes of self-government, with the private
freedom of this or that individual to score his point. "The
primary purpose of the First Amendment," says Meiklejohn,
"is that all the citizens shall, so far as possible, understand the issues that bear upon our common life. That is
why no idea, no opinion ... no relevant information may be
kept from them."*

Now, the enquiry of the scholar—the enlargement of human understanding—is supremely public business. It is the
sovereign exercise of the liberty asserted in the First
Amendment. Free enquiry has immense practical consequences
when it is turned to promoting useful knowledge. In the
words of Thomas Jefferson, "As long as we may think as we
will and speak as we think, the condition of man will proceed in improvement." After two centuries, we can see how
improvement in the condition of man has, in turn, extended
citizenship to the entire population, including the descen-
dants of Jefferson's slaves. Acquantance with this history
should teach us that the scientific enterprise is concerned
not alone with means but with ends and with the framing of
human value and purpose.

The proposal that limits somehow be set upon scientific enquiry must be understood, therefore, as calling for
limits upon the public freedom of citizenship. It requires
a surrender of sovereignty to some agency of the government
which, to that extent, the citizenry would cease to govern.
The extension of the balancing doctrine of clear and present
danger to scientific inquiry strikes at the practice as well
as the principle of self-government.

It is no coincidence that the scientific and democratic
revolution have come forward in history together. A scientist can recognize no authority but his own judgement. On
the choice of his question, the design of the experiment and
the evaluation of the evidence, he must make his own independent determinations. This kind of work, in the words of Per-
cy Bridgman, is "as private as my toothache." The autonomy
of the scientist gives pragmatic sanction to the sanctity of
the individual that makes us all free and equal. Histori-
cally, the archetype is Galileo writing and publishing, in
defiance of the spiritual and temporal authorities who had
placed him under house arrest, the Dialogue on the Two New
Sciences that was to bring those authorities down.

---

*A. Meiklejohn, Political Freedom (NY: Harper and Bros., 1960):
75. Meiklejohn includes in his book a discussion of Justice
Holmes' interpretation of the First Amendment.

Here we confront another paradox. The highly individu-
alistic enterprise of science is, at the same time, an intensely social enterprise. The public nature of the work is
demonstrated by the identity of the "doing" of it with its
publication. Work in science does not win verification or
even the distinction of disproof until it has challenged the
interest of the community.

Science is not, in the phrase of J. Bronowski, a "loose-
leaf notebook." Talk of an "information explosion" totally
misses the true nature of the enterprise; that is the em-
bracing of ever larger and smaller reaches of nature in a
single, connected web of understanding. The worth of a scientist's work is measured by the degree to which it stresses
and reorders the context in which it is done, illuminating
the work of predecessors, contemporaries and successors in
his field.

Thus, despite the efforts of General Leslie Groves to
compartmentalize the Manhattan Project in the name of "security," no major advances in the project were accomplished
out of the knowledge of all participants and even of scientists outside. They could read one another's minds because
they shared a common understanding of the context. The official computation of the chance that the bomb might cata-
lyze the combustion of the atmosphere to nitrous oxide was not
the only one. The same possibility crossed many minds. The
conclusion that the chance was infinitesimal was ballasted in
a redundancy of independent computations. While the chance
that a cabal of crazy geniuses could have otherwise carried
out the experiment was perhaps not equally infinitesimal, it
was securely contained in the openness of the community. The
moratorium called on the recombinant DNA work proceeded from
the same sharing of the context by molecular biologists.

Against fears of nameless evils issuing from the scientific enterprise, therefore, we should place greater confidence in the social constitution of the enterprise than in
limits that might be imposed by some external agency. Attempts to place science under external controls provide, in
fact, recent historic instances of the kind of tyranny against which the scientific enterprise stands. Thus, the
Nazi "doctors" were agents of a pathological dictatorship
that had destroyed the autonomy of German science. They are
convicted of bestiality by the entire irrelevance of their
so-called experiments to the context of work in human biology that has in recent years so totally transformed man's
comprehension of his identity. Similarly, the establishment of the Lysenko lordship over agricultural research in
the U.S.S.R. by the ignorant will of Josef Stalin eradi-

cated a significant line of work in genetics and set that
country's agriculture on the regressive course that has
brought the disastrous crop failures of the 1970s.

In our country today, those participatorily democratic
persons and organizations that presume to speak for the "public" and to call for regulation and restraint of scientific
enquiry must recognize they are tampering with the moral and
pragmatic foundations of our freedom and welfare. Their sur-
prise upon their belated discovery of supreme issues in scientific research reflects their failure as citizens hitherto
to keep themselves informed on important public business. If
they are to contribute constructively to science policy they
must cure themselves of the defect in their intellection that
refers questions of truth to science and questions of value
to other increasingly unspecifiable authority.

Scientists are also citizens. They are the members of
the public best qualified to frame science policy. On questions respecting the hazards that may attend research, they
will necessarily be the first to recognize such hazards and
to provide society's first line of defense against them. With
respect to the propriety of research enterprises, the open
polity of science provides the surest institutional restraint
upon irresponsible or reprehensible individuals. Scientists
have the obligation to inform their fellow citizens of the
nature and the implications of their work; to conduct their
deliberations in the full view of their fellow citizens, and
to invite the participation of those citizens who take the
trouble to make themselves responsibly informed. In the
present controversy over recombinant DNA, the scientific community has given us a practical demonstration of the moral
responsibility that is imposed by the nature of its work and
enforced by its habit of self-government.

Our society already imposes limits upon science in a
host of direct and indirect ways, at serious cost to its own
welfare and to the advance of human understanding. The public funding of science in the name of national defense and
then of national prestige (in the space spectaculars) and now
of cancer cannot have helped but narrow the deployment of the
country's scientific talent within what might otherwise have
been the range of freely elected research ventures. The
present curtailment of public funding is denying the opportunity of graduate education to many young men and women as
fully qualified as those who were getting their doctorates 5
and 10 years ago. As the grim old adage says, we shall always regret most that which we left undone.

Popular anxiety about science would be better directed

at lifting rather than imposing limits. The promotion of the
scientific enterprise and the realization of its benefits to
society invite the responsible participation of all citizens.
With such participation the public funding of science would
not be left to depend upon the ulterior motives of "mission-
oriented"—meaning principally military and medical—agencies.
Nor would decision as to the application of new technology be
abdicated to the private governments of a few hundred major
corporations and the increasingly secretive managements of a
few public agencies.

Happily, there are signs that our resilient democracy is
making progress in this regard. Through such novel institutions as the Environmental Protection Agency and the new Office of Technology Assessment, informed and concerned citizens are indeed taking a hand in the making of public policy
and decision about science and technology. There is nothing
exclusive about the scientific community. One day perhaps it
will enroll the entire citizenry. Openness, reason and tolerance are qualities becoming to the behavior of everyone.

6

The Presumptions of Science

Robert L. Sinsheimer

Can there be "forbidden"--or, as I prefer, "inopportune" knowledge? Could there be knowledge, the possession of which, at a given time and stage of social development, would be inimical to human welfare--and even fatal to the further accumulation of knowledge? Could it be that just as the information latent in the genome of a developing organism must be revealed in an orderly pattern, else disaster ensue, so must our knowledge of the universe be acquired in a measured order, else disaster ensue?

Biological organisms are equipped with many sensors essential to their survival, sensors for heat, cold, pain, thirst, hunger. Social organisms similarly need sensors of peril, particularly as they evolve into new domains--and for these we must use our intelligence, limited as it may be.

Discussion of the possible restraint of inquiry touches a most sensitive nerve in the academic community. If one believes that the highest purpose available to humanity is the acquisition of knowledge (and in particular of scientific knowledge, knowledge of the natural universe), then one will regard any attempt to limit or direct the search for knowledge as deplorable--or worse.

If, however, one believes that there may be other values to be held even higher than the acquisition of knowledge--for instance, general human welfare--and that science and possible other modes of knowledge acquisition should subserve these higher values, then one is willing to (indeed, one must) consider such issues as: the possible restriction of the rate of acquisition of scientific knowledge to an "optimal   level relative to the social context into which

Reprinted by permission of DAEDALUS, Journal of the American Academy of Arts and Sciences, Boston, MA. Spring 1978, Limits of Scientific Inquiry.

it is brought; the selection of certain areas of scientific research as more or less appropriate for that social context; the relative priorities at a given time of the acquisition of scientific knowledge or of other knowledge such as the effectiveness of modes of social integration, or of systems of justice, or of educational patterns.

In short, if one does not regard the acquisition of scientific knowledge as an unquestioned ultimate good, one is willing to consider its disciplined direction. One may, of course, still have grave doubt as to whether mankind can know enough to be able intelligently to guide the rate or direction of the scientific endeavor, but at least one will then accept that we have a responsibility to seek answers-- if there be any--to such questions.

The Impact of Science

In 1930 Robert A. Millikan, Nobel Prize winner, founder and long-time leader of Caltech, wrote in an article entitled "The Alleged Sins of Science" that one may

> sleep in peace with the consciousness that the Creator has put some foolproof elements into his handiwork, and that man is powerless to do it any titanic physical damage.[1]

To what was Millikan referring? Stimulated by the recombinant DNA controversy, I have looked back to see if there were any similar admonitions or premonitions with respect to the possible consequences of nuclear energy. And there were. Millikan, in 1930, was responding to an earlier writing of Frederick Soddy. In a book entitled Science and Life Soddy, who had been a collaborator of Rutherford, had written:

> Let us suppose that it became possible to extract the energy which now oozes out, so to speak, from radioactive material over a period of thousands of millions of years, in as short a time as we pleased. From a pound weight of such substance one could get about as much energy as would be obtained by burning 150 tons of coal. How splendid. Or a pound weight could be made to do the work of 150 tons of dynamite. Ah, there's the rub . . . It is a discovery that conceivably might be made tomorrow in time for its development and perfection, for the use or destruction, let us say, of the next generations, and, which it is pretty certain, will be made by science

sooner or later. Surely it will not need this
actual demonstration to convince the world that
it is doomed if it fools with the achievements
of science as it has fooled too long in the
past.

War, unless in the meantime man has found a
better use for the gifts of science, would not
be the lingering agony it is today. Any selected
section of the world, or the whole of it if
necessary, could be depopulated with a swift-
ness and dispatch that would leave nothing to
be desired.[2]

Millikan commented, just prior to his statement quoted above,
"Since Mr. Soddy raised the hobgoblin of dangerous quantities
of available subatomic energy [science] has brought to light
good evidence that this particular hobgoblin--like most of
the hobgoblins that crowd in on the mind of ignorance--was a
myth . . . The new evidence born of further scientific
study is to the effect that it is highly improbable that
there is any appreciable amount of available subatomic
energy to tap."[3]

So much for scientific prophecy. But it is indeed
instructive and also troubling to recognize that our scientific endeavor truly does rest upon unspoken, even unrecog-
nized, faith--a faith in the resilience, even the benevol-
ence, of nature as we have probed it, dissected it,
rearranged its components in novel configurations, bent its
forms and diverted its forces to human purpose. Scientific
endeavor rests upon the faith that our scientific probing
and our technological ventures will not displace some key
element of our protective environment and thereby collapse
our ecological niche. It is a faith that nature does not
set booby traps for unwary species.

Our bold scientific thrust into new territories un-
charted by experiment and unencompassed by theory must rely
wholly upon our faith in the resilience of nature. In the
past that faith has been justified and rewarded, but will
it always be so? The faith of one era is not always appropriate to the next, and an unexamined faith is unworthy of
science. Ought we step more cautiously as we explore the
deeper levels of matter and life?

Most states of nature are quasiequilibria, the outcome
of competing forces. Small deviations from equilibrium, the
result of natural processes or human intervention, are most
often countered by an opposing force and the equilibrium

restored, at some rate dependent upon the kinetics of the processes, the sizes of the relevant natural pools of components, and other factors. Although we may therefore speak, of the resilience of nature, this restorative capacity is finite and is limited in rate.

For example, if the ozone layer of the atmosphere is lightly and transiently depleted by a nuclear explosion or the atmospheric release of fluorocarbons, the natural processes which generate the ozone layer can restore it to the original level within a brief period. However, should the ozone layer be massively depleted--as by extended, large-scale release of fluorocarbons--many decades would be required for its renewal by natural processes, even if the release of fluorocarbons ceased.

· Similarly, the populations of most living creatures can achieve an equilibrium level dependent upon birth rates and upon death rates from various causes. Most species have an excess capacity for reproduction, so that minor additions to the process of their removal (as by the harvesting of fish) cannot appreciably influence the equilibrium population. Patently however, excessive harvesting removing numbers beyond the reproductive capacity of the species will in time bring about its extinction.

In a similar manner lakes and rivers and air basins can absorb and dispose of limited amounts of pollutant but can be overwhelmed by masses beyond their capacity. Once overwhelmed the very agents responsible for disposal of pollution in small quantities may be destroyed, leaving a "dead" sea.

The concept of resilience extends to the planet as a whole and to the impact upon the manifold equilibria upon which the network of life forms depends as we continue to expand our intensive monoculture agriculture, as we continue to increase the total of human energy consumption (the man-made release of energy in the Los Angeles basin is now estimated at about 5 percent of the solar input), as we continue to raise the atmospheric level of CO_2 by combustion of fossil fuel, and so forth.

Because human beings (and most creatures) are adapted by evolution to the near equilibrium states, the resilience provided by the restorative forces of nature has appeared to us to be not only benevolent, but unalterable. Less overt than our faith in the resilience of nature is the faith with which we have relied upon the resilience of our social institutions and their capacity to contain the stress of change and to adapt the knowledge gained by science--and the power inherent in that knowledge--to the benefit of society, more than to its detriment. The fragility of the equilibria underlying social institutions is even more apparent than of the equilibria of nature. Political, economic, and cultural balances have shifted drastically in human history under the impact of new technologies, of new ideologies or religions, of invading peoples, of resource exhaustion, and other changes. Our faith in the resilience of both natural and man-made phenomena is increasingly strained by the acceleration of technical change and the magnitude of the powers deployed.

Physics and chemistry have given us the power to reshape the physical nature of the planet. We wield forces comparable to, even greater than those of, natural catastrophes. And now biology is bringing to us a comparable power over the world of life. The recombinant DNA technology, while significant and potentially a grievous hazard in itself (through the conceivable production, by design or by inadvertance, of new human, animal, or plant pathogens or of novel forms capable of disrupting important biological equilibria), must be seen as a portent of things to come.

The present recombinant DNA technology, which permits the addition or replacement of a few genes in living cells, is but the first prototype of genetic engineering. More powerful means involving cell fusion or chromosome transfer are already close to hand; even more sophisticated future developments appear assured. Since genes determine the basic structures and biological potentials of all living forms, the ultimate potential of genetic engineering for the modification and redesign of plants and animals to meet human needs and desires seems virtually unlimited.

Such capabilities will pose major questions as to the extent to which mankind will want to assume the responsibility for the life forms of the planet. Further, there is no reason to believe the same technology will not be applicable to mankind as well, the capability of human genetic engineering will raise profound questions of values and judgment for human societies.

It seems paradoxical that a living organism emergent from the evolutionary process after billions of years of blind circumstance should undertake to determine its own future evolution. The process is perhaps analogous to that of the mind seeking to understand itself. In both cases it is uncertain whether the attempt can possibly be successful. Nonetheless, at this point perhaps we had best step back and reconsider what it is we are about.

For four centuries science has progressively expanded
our knowledge and reshaped our perception of the world. In
that same time technology has correspondingly reshaped the
pattern of our lives and the world in which we live.

Most people would agree that the net consequence of
these activities has been benign. But it may be that the
conditions which fostered such a benign outcome of scientific advance and technological innovation are changing to a
less favorable set. Changes in the nature of science or
technology or in the external society--in either the scale
of events or their temporal order--can affect the precon-
ditions, the presumptions, of scientific activity, and can
thus alter the future consequences of such activities.

Both quantitative and qualitative changes have surely
affected the impact of science and technology upon society.
Quantitatively, the exponential growth of scientific
activity and the unprecedented magnitude of modern industrial ventures permit the introduction of new technologies
(e.g., fluorocarbon sprays) on a massive scale within very
brief periods often with unforeseen consequence. Qualitatively, science and technology have been directed increasingly to synthesis--to the formulation of new substances
designed for specific human purpose. Thus we have synthetic
atoms (plutonium, strontium-90), synthetic molecules
(dioxin, kepone, DDT) and now synthetic microorganisms
(recombinant DNA). In these activities we introduce wholly
novel substances into the planetary environment, substances
with which our evolution has not always prepared us to cope.

Can we continue to rely upon the past four centuries as
a guide for scientific activity, given these changes? Other
human activities of this same era are now increasingly seen
in a different hue. The same period witnesses exponential
increases in population and in the exploitation of natural
resources for material wealth. Few would argue continuance
of such trends will be benign.[4] The same era has witnessed
the constant acceleration of the rate of change, the
increasing dominance of technology in the affairs of men.

The constantly accelerating accretion of knowledge,
therefore, may not always be counted as a good. Can circumstances change so as to devalue the net worth of new knowledge? Might a pause or slowdown for consolidation and
reflection then be more in order? Indeed, could it be that
some knowledge could, at this time, be positively malign?
Hard questions, perhaps not answerable, perhaps not the
right questions, but they are not answered for 1977 by
invoking Galileo or Darwin or Freud. I believe they demand

our thought.

I would advance for consideration some propositions
that frankly I'm not at all sure I entirely believe. I
think that in order to find out what one does believe it is
necessary to go beyond what one can readily accept--to
explore honestly more extreme and more remote positions so
that one's position is based upon intelligent choice, not
simple ignorance.

The domain I propose to explore can be indicated by a
question. The question is one I have actually raised within
the administration at Caltech (and it could as well be
raised elsewhere). Institutions such as Caltech and others
devote much energy and effort and talent to the advancement
of science. We raise funds, we provide laboratories, we
train students, and so on. In so doing we apply essentially
only one criterion--that it be good science as science--that
the work be imaginative, skillfully done, in the forefront
of the field. Is that, as we approach the end of the
twentieth century, enough? As social institutions, do Caltech and others have an obligation to be concerned about
the likely <u>consequences</u> of the research they foster? And
if so, how might they implement such a responsibility?

For reasons which probably need no elaboration Caltech
has been more than reluctant to come to grips with this
question. And, indeed, it just may be--and I say this with
real sorrow--that scientists are simply not the people
qualified to cope with such a question. The basic tactic
of natural science is analysis: fragment a phenomenon into
its components, analyze each part and process in isolation,
and thereby derive an understanding of the subject. In
physics, chemistry, even biology, this tactic has worked
splendidly.

To answer my question, however, the focus must not be
inward but outward, not narrowed but broadened. The focus
must be on all the ties of the sciences to society and
culture and on the impact of scientific knowledge and
technological advancement on all human, indeed all
planetary, life.

Consider as an instance the recombinant DNA issue. The
natural tendency of the scientist, if he will admit this a
problem, is to break it down, to decompose it into individually analyzable situations. If there is a danger, quan-
titate it: what is the numerical chance of the organisms
escaping, of their colonizing the gut, of their penetrating
the intestinal epithelium, of their causing disease (what

disease)? If you point out that there is a nearly infinite
set of possible scenarios of misfortune--that accidents do
happen and in unpredictable ways, that humans do err, that
bacterial or viral cultures do become contaminated, that
indeed aspects of this technology involve inherently unpredictable consequence and hence are not susceptible to
quantitative analysis--you are regarded as unscientific.

The consequences of the interaction of known but
foreign gene products with the complex contents of a bacterial cell would be difficult enough to predict, much less
the consequences of the interactions of unknown gene products, as produced in "shotgun" experiments. Some of these
consequences may well modify, in unpredictable ways, the
likelihood of the organism's survival or persistence in
various environments, its potential toxicity for a host or
nearby life forms. It may alter, for instance, an organism's
survival in an animal intestine, contrary to our expectations, for we have presumed that we know all factors important for survival there and that no new successful adaptations could emerge.

For complex reasons, consideration of the potential
hazards from organisms with recombinant DNA has focused upon
immediate medical concerns. That these organisms with
unpredictable properties might have impact upon any of the
numerous microbiological processes which are important components of our life support systems is simply dismissed as
improbable. The fact that these organisms are evolutionary
innovations and have within themselves, as do all living
forms, the capacity (if they survive) for their own unpredictable future evolutionary development is ignored, or
dismissed as mystical.

If you point out that the recombinant DNA issue simply
cannot be effectively considered in isolation but must be
viewed in perspective and in a larger context as a possible
precursor to future technologies available to many elements
of society (including totalitarian governments, the military, and terrorist factions) your remarks are regarded as
irrelevant to science.

There is an intensity of focus in the scientific perspective which is both its immediate strength and its ultimate weakness. The scientific approach focuses rigorously
upon the problem at hand, ignoring as irrelevant the ante-
cedents or motive and the prospectives of consequence.

Viewed objectively such an approach can only make sense
if either (1) the consequences are always trivial, which is

patently untrue, or (2) the consequences are always benign, that is, if the acquisition of knowledge, of any knowledge at any time, is always good, a proposition one might find hard to defend, or (3) the dangers and difficulties inherent in any attempt to restrict the acquisition of knowledge are so great as to make the unhindered pursuit of science the lesser evil.

In thinking about the impacts of science, we should, perhaps, reflect upon the inverse of the uncertainty principle. Perhaps it might be called the certainty principle. The uncertainty principle is concerned with the inevitable impact of the observer upon the observed, which thereby alters the observed. Conversely, there is an effect of the observed upon the observer. The discovery of new knowledge, the addition of new certainty, which correspondingly diminishes the domain of uncertainty and mystery, inevitably alters the perspective of the observer. We do not see the world with the same eyes as a Newton or a Descartes, or even a Faraday or a Rutherford.

The acquisition of a discipline sharpens our vision in its domain, but too frequently it seems also to blind us to other concerns. Thus immersion in the world of science, with its store of accumulated and substantiated fact, can make the participant intolerant of, and impatient with the uncertainties and non-reproducibilities of the human world. Engrossed in the search for knowledge, scientists tend to adopt the position that more knowledge is the key to the solution to human problems. They may not see that the uses we make of knowledge or the ways in which we organize to use knowledge can, as well, be the limiting factors to the human condition, and they forget that even within science our knowledge and our theories are always human constructs. Moreover, we should always remember (lest we become too secure and even smug) that our knowledge and our theories are ever incomplete.

Of Dubious Merit

To make this discussion more specific let me consider three examples of research that I personally consider to be, on balance, of dubious merit. One is in an area of rather applied research, the second in a very speculative but surely basic area, and the third in the domain of biomedical research, which we most often conceive to be wholly benign.

The first I would cite is current research upon improved means for isotope fractionation. In one technique, one attempts to use sophisticated lasers[5] to activate

selectively one isotope of a set. I do not wish to discuss
the technology but rather the likely consequence of its
success. To be sure, there are benign experiments that
would be facilitated by the availability of less expensive,
pure isotopes. For some years I wanted to do an experiment
with oxygen 18 but was always deterred by the cost.

But does anyone doubt that the most immediate application of isotope fractionation techniques would be the separation of uranium isotopes? This country has recently
chosen to defer, at least, if not in fact to abandon, the
plutonium economy and the breeder reactor because of well-
founded concern that plutonium would inevitably find its way
into weapons. We are thus left with uranium-fueled reactors.
But uranium 235 can also be made into a bomb. Its use for
power is safer only because of the difficulty in the separation of uranium 235 from the more abundant uranium 238. If
we supersede the complex technology of Oak Ridge, if we
devise quick and ingenious means for isotope separation,
then one of the last defenses against nuclear terror will
be breached. Is the advantage worth the price?

A second instance I would cite of research of dubious
merit, and one probably even more tendentious than the
first, relates to the proposal to search for and contact
extraterrestrial intelligence.[6] Recent proposals suggest
that, using advanced electronic and computer technology, we
could monitor a million "channels" in a likely region of the
electromagnetic spectrum, "listening" over several years
for signals with an "unnatural" regularity or complexity.

I am concerned about the psychological impact upon
humanity of such contact. We have had the technical
capacity to search for such postulated intelligence for less
than two decades, an instant in cosmic terms. If such
intelligent societies exist and if we can "hear" them, we
are almost certain to be technologically less advanced and
thus distinctly inferior in our development to theirs. What
would be the impact of such knowledge upon human values?

Copernicus was a deep cultural shock to man. The
universe did not revolve about us. But God works in
mysterious ways and we could still be at the center of
importance in His universe. Darwin was a deep cultural
shock to man. But we were still number one. If we are
closer to the animals than we thought before, and through
them to the rocks and the sea, it does not really devalue
man to revalue matter. To really be number two, or number
37, or in truth to be wholly outclassed, an inferior species,
inferior on our own turf of intellect and creativity and

imagination, would, I think, be very hard for humanity.

The impact of more advanced cultures upon less advanced has almost invariably been disastrous to the latter. We are well acquainted with such impacts as the Spanish upon the Aztecs and Incas or the British and French upon the Polynesians and Hawaiians. These instances were, however, compounded by physical interventions (warfare) and the introduction of novel diseases. I want to emphasize the purely cultural shock. Hard learned skills determinant of social usefulness and positions become quickly obsolete. Less advanced cultures quickly become derivative, seeking technological handouts. What would happen to our essential tradition of self-reliance? Would we be reduced to seekers of cosmic handouts?

The distance of the contacted society might, to some degree, mitigate its consequent impact. A contact with a round trip communication time of ten years would have much more effect than one with a thousand years. The likelihood of either is, however, a priori, unknown. Nor is it inconceivable that an advanced society could devise means for communication faster than light.

The proponents of such interactions have considered the consequences briefly. In a 427-page book Communication with Extraterrestrial Intelligence[7] sixteen pages comprise a chapter entitled "Consequences of Contact." Opinion therein ranges from "Our obligation is, I feel, to stress that in any sensible way this problem has no danger for human society. I believe we can give a full guarantee of this" to "If we come in contact with some superior civilization this would mean the end of our civilization, although that might take a while. Our period of culture would be finished."

How and by whom should such a momentous decision[8] be made--one that will clearly, if successful, have an impact upon all humanity? Somehow I cannot believe it should be left to a small group of enthusiastic radioastronomers.

My concern here does not extend so far that I would abolish the science of astronomy. If the astronomers in the course of their science come across phenomena that can only be understood as the product of intelligent activity, so be it. But I do not believe that is the same as deliberately setting out to look for such activity with overt pretensions of social benefit.

The third example of research I consider of dubious merit concerns the aging process. I would suggest this

subject exemplifies in supreme degree the eternal conflict
between the welfare of the individual and the welfare of
society and, indeed, the species. Obviously, as individuals,
we would prefer youth and continued life. Equally obviously,
on a finite planet, extended individual life must restrict
the production of new individuals and that renewal which
provides the vitality of our species.

The logic is inexorable. In a finite world the end of
death means the end of birth. Who will be the last born?

If we propose such research we must take seriously the
possibility of its success. The impact of a major extension
of the human life span upon our entire social order, upon
the life styles, mores, and adaptations associated with
"three scores and ten," upon the carrying capacity of a
planet already facing over-population would be devastating.
At this time we hardly need such enormous additional problems. Research on aging seems to me to exemplify the
wrong research on the wrong problem in the wrong era. We
need that talent elsewhere.

Is Restraint Feasible?

If one concedes, however reluctantly, that restraint
of some directions of scientific inquiry is desirable, it is
appropriate to ask if it is feasible and, if so, at what
cost.

Some of my colleagues not only in biology but in other
fields of science as well, have indicated to me that they
too increasingly sense that our curiosity, our exploration
of nature, may unwittingly lead us into an irretrievable
disaster. But they argue we have no alternative.[9] Such a
position is, of course, a self-fulfilling prophecy.

I would differentiate among what might be called
physical feasibility, logical feasibility, and political
feasibility.

I believe that actual physical restraint is in principle feasible. There are two evident avenues of control:
the power of the purse and access to instruments. Control
of funding is indeed already a powerful means for control
of the directions of inquiry for better or worse. To the
extent that there exists a multiplicity of sources of
support, such control is porous and incomplete, but it is
clearly a first line of restraint.

Research today cannot be done with household tools. It is difficult to imagine, for instance, any serious research on aging that would not require the use of radioisotopes or an ultracentrifuge or an electron microscope. The use of isotopes is already regulated for other reasons. Access to electron microscopes could, in principle, be regulated, albeit at very real cost to our current concepts of intellectual freedom.

An immediately related, important aspect of any policy of restraint concerns the distinctions to be made about the nature of research. Can we logically differentiate research on aging from general basic biologic studies? I expect we cannot in any simple, absolute sense. Yet obviously the people who established the National Institute of Aging must have believed that there is a class of studies which deserves specific support under that rubric. Indeed, distinctions of this sort are made all the time by the various institutes of National Institutes of Health in deciding which grant applications are potentially eligible for their particular support. Pragmatically, and with some considerable margin of error, such distinctions can be and are made.

It is frequently claimed that the "unpredictability" of the outcome of research makes its restraint, for social or other purpose, illogical and indeed futile. However, the unpredictability of a research outcome is not an absolute but is both quantitatively and qualitatively variable.

In more applied research within a field with well-defined general principles, the range of possible outcomes is surely circumscribed. In more fundamental research, in wholly new fields remote from prior human experience--as in the cosmos, or the subatomic world, or the core of the planet--wholly novel phenomena may be discovered. But, for instance, even in a fundamental science such as biology, most of the overt phenomena of life have been long known.

The basic principles of heredity were discovered by Mendel a century ago and were elaborated by Morgan and others early in this century. The understanding of genetic mechanism, the reduction of genetics to chemistry, had to await the advent of molecular biology. This understanding of mechanism has now provided the potential for human interventions, for genetic engineering, but it has not significantly modified our comprehension of the genetic basis of biological process.[10]

The path of modern biology will surely lead to further

understanding of biological mechanism, with subsequent
application to medicine and agriculture (and accompanying
social impact). But it would seem likely that only within
the central nervous system may there be the potential for
wholly novel--and correspondingly wholly unpredictable--
process. Even there, the facts of human psychology and the
subjective realities of human consciousness have long been
familiar to us, albeit the underlying mechanisms are indeed
obscure.

Political feasibility is, of course, another question.
The constituency most immediately affected is, of course,
the scientific. And despite our protestations and alarms
this community does have real political influence. It would
seem unlikely to me that a policy of scientific restraint
could be adopted in any sector unless a major portion of the
scientific community came to believe it desirable.

For this to happen, that community will clearly have to
become far more alert to, and aware of, and responsible for
the consequences of their activities. The best discipline
is self-discipline. Scientists are keenly sensitive to the
evaluations of their peers. The scientific community and
the leaders of our scientific and technical institutions
will have to develop a collective conscience; they will have
to let it be known certain types of research are looked upon
askance, much as biological warfare research is today; it
needs to be understood that such research will not be
weighed in considerations of tenure and promotion; societies need to agree not to sponsor symposia on such topics.
All of these and similar measures short of law could indeed
be very effective.

I am well aware of the dangers implicit in such forms
of cultural restraint. But I think we really must look at
the dangers we face in the absence of self-restraint. Do
we accept only the restraint of catastrophe?

If we are to consider this position, we must do so in
a forthright manner. We must be willing to explore the
vistas exposed if we lower conventional taboos and sanctions. We may not at first enjoy what we see, but at
least we will have a better perception of the available
alternatives. Any attempt to limit the freedom of scientific inquiry will surely involve what will appear, at
least at first, to be quite arbitrary distinctions--
judgmental decisions, the establishment of boundaries in
gray and amorphous terrain. These are, however, familiar
processes in our society, in the courts, in the legislatures. Indeed, most of us are familiar with such

problems in our educational activities. The selection of
new faculty, the award of tenure, the assignment of grades
are clearly judgmental decisions.

In science we try with some success to elude the
necessity for such very human judgments. Indeed, one sus-
pects that many persons go into science precisely to avoid
the necessity for such complex decisions--in search of a
domain of unique and unequivocal answers of enduring
validity. And it is painful to see the sanctuary invaded.

Admittedly it is difficult to achieve consensus on the
criteria for judgmental decisions. Such consensus is all
the more difficult in the sphere of international activities
such as science which involve participants from diverse
cultures and traditions.

Conversely there are many persons who prefer the more
common, perhaps the more human world of ambiguity and compromise and temporally valid judgments and who resist the
seemingly brutal, life and death, cataclysmic types of
decision increasingly imposed upon society by the works of
science. And science and scientists cannot stand wholly
aloof from the latter dilemmas--for science is a human
activity and scientists live in the human society. We cannot expect the adaptation to be wholly one-sided.

Even if, at best, we can only slow the rate of
acquisition of certain areas of knowledge, such a tactic
would give us more time to prepare for social adaptation--
if we mobilize ourselves to use that time.

The Case for Restraint

The view one exposes by lifting that sanction we label
freedom of inquiry is frankly gloomy. It would seem that we
are asked to make thorny decisions and delicate differentia-
tions, to relinquish long-cherished rights of free inquiry,
to forego clear prospects of technological progress. And
it would seem that all these concessions stem ultimately
from recognition of human frailty and from recognition of
the limitations of human rationality and foresight, of
human adaptability and even good will. Just such recogni-
tions have already spawned many of our institutions and
professions--religions, the law, government, United
Nations--yet all of these are as imperfect as the world
they are designed to restrain and improve.

At each level of human activity, whether individual,
group, or national, we continually struggle to find

acceptable compromises between the freedom to pursue varied courses and goals and the conflicts that arise when one person's actions run contrary to another's. In a crude sense the greater the power available to an entity, the more limitations must be imposed upon its freedom if conflict is to be averted. Ideally such limits are internalized through education and conscience, but we all understand the inadequacy of that process.

In short, we must pay a price for freedom, for the toleration of diversity, even eccentricity. That price may require that we forego certain technologies, even certain lines of inquiry where the likely application is incompatible with the maintenance of other freedoms. If this is so and if we can recognize and understand this, perhaps we can, as scientists, be more accepting.

Some will argue that knowledge simply provides us with more options and thus that the decision point should not be at the acquisition of knowledge but at its application.

Such a view, however ideal, overlooks the difficulty inherent in the restriction of application of new knowledge, once that knowledge has become available in a free society. Does anyone really believe, for instance, that knowledge permitting an extension of the human life span would not be applied once it were available?

One must also recognize again that the very acquisition of knowledge can change both the perceptions and the values of the acquirer. Could, for instance, deeper knowledge of the realities of human genetics affect our commitment to democracy?

It may be argued that the cost, however it may be measured, of impeding research would be greater to a society than the cost of impeding application. Perhaps so. This issue could be debated, but it must be debated in realistic terms with regard for the nature of real people and real society and with full understanding that knowledge is indeed power.

Although the nature of the measures necessary to restrict the application of knowledge has seldom been analyzed, the measures needed would surely be dependent upon the size of investment required to apply the knowledge, as well as on the form of and the need for the potential benefits of the knowledge, among other things. The compatibil-

ity of such restrictive measures with the principles of a
democratic society would need to be considered. Restriction of nuclear power may be a case in point.

Alvin Weinberg has developed the concept of the technological fix as the simple solution to cut the Gordian knot
of complex social problems. However, we seem to be discovering that the application of one technological fix seems
to lead us into another technological fix. For example,
the development of antibiotics and other triumphs of modern
medicine has led to the tyranny of overpopulation. In
efforts to cope with overpopulation by more intensive
agriculture, we develop pesticides, herbicides and other
chemicals which increase the level of environmental
carcinogenesis. And so on.

The moral is that we cannot ignore the social and
cultural context within which the technology is deployed.
In retrospect we can see that in the cultural and social
context of the seventeenth, eighteenth, and nineteenth
centuries the consequences of technological innovation
were most often benign. Whether because of change in the
society and culture or change in the nature and effectiveness of technology, at some time in the twentieth century
the balance began to shift and by now our addiction to
technology begins to assume an unpleasant cast.

We are indeed addicted to technology. We rely ever
more upon it and thus become its servant as well as its
master. It has led to human populations insupportable
without its aid. Further, new technologies shape our
perceptions; they spawn expectations of change or stir
deep fears of disaster. They dissociate us from the past
and becloud the shape of the future. Even the oldest
boundary conditions of humanity fall as we leave the
planet and as we plan to reshape our genes.

Our academic institutions and our professional societies foster and promote science. To some degree they
also have concern for its consequences, but it is a minor
aspect. The principle that one should separate agencies
which promote and agencies which regulate may apply here.

But where then is the balance, the necessary check to
the force of scientific progress? Is the accumulation of
knowledge unique among human activities--an unmitigated
good that needs no counterweight? Perhaps that was true
when science was young and impotent, but hardly now. Yet
we lack the institutional mechanisms for regulation.

Our experience with constraint upon science has hardly been encouraging. From the Inquisition to Lysenko such constraint has been the work of bigots and charlatans. Obviously, if it is to be done to a good purpose, any restraint must be informed, both as to science and as to the larger society on which science impacts.

The acquisition of knowledge is a human, a social, enterprise. If we, through the relentless, single-minded pursuit of new knowledge so destabilize society as to render it incapable—or unwilling—to continue to support the scientific enterprise, then we will have, through our obsession, defeated ourselves.

At Caltech and the many other academic institutions, we have now, <u>culturally</u>, cloned Galileo a millionfold. We have nurtured this Galilean clone well; we award prizes and honors to those most like the original. No doubt this clone has been most beneficial for humanity, but perhaps there is a time for Galileos. Perhaps we need in this time to start another clone.

References

1. R. A. Millikan, "Alleged Sins of Science," <u>Scribners Magazine</u>, 87 (2) (1930): 119-130.

2. Frederick Soddy, <u>Science and Life</u> (London: John Murray, 1920).

3. Precisely what evidence Dr. Millikan had in mind is uncertain. However, it was generally appreciated that the efficiency of nuclear transformation by the charged particles then in use was so low that there was no significant prospect of a net release of energy. No practical chain reaction could yet be envisaged.

4. A. V. Hill in his presidential address to the British Association for the Advancement of Science in 1952, referring to the population problem, said, "If ethical principles deny our right to do evil in order that good may come, are we justified in doing good when the foreseeable consequence is evil?"

5. See A. S. Krass, "Laser Enrichment of Uranium: The Proliferation Connection," <u>Science</u>, 196 (1977): 721-731; also B. M. Casper, "Laser Enrichment: A New Path to Proliferation?" <u>Bulletin of Atomic Scientists</u>, 33 (1) (1977): 28-41.

6. See T. B. H. Kuiper and M. Morris, "Searching for Extraterrestrial Civilizations," Science, 196 (1977): 616-621; also B. Murray, S. Gulkis, and R. E. Edelson, "Extraterrestrial Intelligence: An Observational Approach," Science, in press.

7. C. Sagen (ed.), Communication with Extraterrestrial Intelligence (EETC) (Cambridge, Mass.:MIT Press, 1973).

8. Conceivably, we might not be given this choice if an advanced civilization were determined to contact us. At present however, it would seem to be our option.

9. This is not a new perception. "The world is now faced with a self-evolving system which it cannot stop. There are dangers and advantages in this situation. . . . Modern science has imposed upon humanity the necessity for wandering. Its progressive thought and its progressive technology make the transition through time, from generation to generation, a true migration into unchartered seas of adventure. The very benefit of wandering is that it is dangerous and needs skill to avert evils. We must expect, therefore, that the future will disclose dangers. It is the business of the future to be dangerous; and it is the merit of science that it equips the future for its duties," wrote A. N. Whitehead in Science and the Modern World (New York: Free Press, 1967), 205-07.

10. Indeed the failure to discover a new class of phenomena underlying genetics has been most disappointing to some. See Gunther S. Stent, "That Was the Molecular Biology That Was," Science, 160 (1968): 390-395.

7

Secrecy and the Individual in Sociological Research

Kurt W. Back

The Limits of Knowledge

Science has achieved an important and powerful position in contemporary society. This achievement is witnessed by the fact that social limitations on the scientific enterprise are seen as unduly restrictive and even as the insidious work of opponents of the untrammeled pursuit of knowledge. This complaint may be justified, certainly from the point of view of the scientist. But in chafing under restriction scientists tend to forget how unusual this high value on acquisition of knowledge is in human history and how precarious the status of seeking knowledge is even in our own society. Human history certainly gives no warrant for the belief that untrammeled scientific activity is a normal condition, but rather suggests the opposite belief, namely that there is a need for limits; or it may be that the drive for acquisition of information is in some way balanced with this need. Understanding the nature, conditions and strength of these limits may help us in understanding the current problems surrounding the social control of science.

The move toward unrestrained pursuit of science is a novel development in human history. Its position is intimately connected with the value given to knowledge. The rise in importance of this value has been recent, since the Renaissance in the western world, and acceptance of the importance of disinterested knowledge is still precarious. The importance of the status of the scientist, representing

---

Research reported here was supported by a grant from the Mary Duke Biddle Foundation. The assistance of Marianella Canton, Margaret Rogers and Julia Burchett is gratefully acknowledged.

a value which has been of such subsidiary importance, has
produced a tension in our society which leads to clashes on
the question of the social control of science. Many people
are still dubious of any search for knowledge for its own
sake, thinking it idle curiosity or worse, or they suspect
anyone who does so of an ulterior aim, of searching for power
or of serving sinister forces. The picture of the pure scientist oscillates in the popular mind between that of the
dilettante and the ogre; only when science and knowledge are
shown to serve some other goal, if science is transformed into technology, is the place of the scientist really secure.

Knowledge as a Value

One may disagree with the picture presented here of the
low value placed on the acquisition of knowledge in general
human life. To demonstrate the position of scientific knowledge in our society we can start by comparing it with another similar value, namely beauty and its social expression,
art. Here too we find a gradual emancipation of the artist,
an acceptance of beauty as a good; the enterprise of art,
once considered by many as a frivolous activity or only justified as artisanship in the service of other values, is
gradually being freed from restrictions. However, art has
been accepted in many times and places and if this is not
the only period which gives a special place to the artist,
it is almost unique in doing so, however grudgingly, to the
scientist. Still, art seems to be more securely accepted: a
slogan of science for science's sake sounds more peculiar
than art for art's sake.

We can go even further and note that knowledge has some
dangerous, fear-inspiring connotations. Looking at myths as
representations of ancient world-views, at the origins of
the two main myths on which Western society rests, Jewish
and Greek, we find that both put desire for knowledge as the
base of all human woes, symbolized by the apple from the
Tree of Knowledge and Pandora's box. In myth after myth we
find this theme repeated, that wanting some information will
lead to disaster; in the stories of Oedipus, of Lohengrin,
up to the mad scientist such as Dr. Frankenstein, looking
for knowledge leads to a bad end. At the beginnings of modern science, we have the cautionary tale of Faust, who sold
his soul to the devil for knowledge. And he is based on
Paracelsus, one of the earliest scientists. It seems that
whenever a limit is set on knowledge this becomes almost a
signal for the audience that it will be transgressed with
horrible results, whether the protagonist is Eve, Blue-
beard's wife, or Elsa of Brabant.

We can readily see that limitations on science will occur when scientific pursuits clash with other values. However, as we have proposed, science and knowledge have had a
precarious position throughout most of the history of the human race and we should not be surprised if the limitations on
science are readily accepted, or even eagerly sought. Studies on the values of our society will confirm this predic-
tion.

Even in our presumably scientific society there does not
seem to be any general appreciation of science's intrinsic
value. Studies of values frequently do not even include
knowledge. Value orientations do not give the ideal of pure
understanding a particularly high place. In Spranger's study
of human orientation (7) there is a place for 'theoretical
man' whose aim is to understand the nature of things. The
Allport-Vernon Study of Values (1) is based on Spranger's ty-
pology and is one of the most widely used tests of values,
and it takes theoretical values as one of its orientations.
In other more recent tests, however, knowledge fared less
well. Charles Morris (5) identified a number of value com-
plexes, which he called paths of life, naming each path either after its representation in a culture or after a divin-
ity. But in his system there is no Athenian path, looking
for enlightenment; he shows the Promethean path, which is not
looking for knowledge but for technology. By contrast, we
find an Apollonian path, looking for harmony and beauty.
Perhaps a little less esoteric is the recent study of Milton
Rokeach (6) based on intensive review of the literature in
empirical research. He has twenty ultimate and thirteen in-
strumental values. Knowledge finds no place amongst the ultimate values at all, though wisdom does; the closest instru-
mental value is intellectuality, which is typically ranked
very low. The work of psychologists who study human values
individually confirms the low appreciation of science and
knowledge for its own sake.

Under these conditions we should not be surprised that
scientific endeavor is subject to strong pressures. The situation, however, is worse than that. In a very fundamental
way knowledge is seen not as a blessing but as a sinister
threat. We have seen the degradation of the search for knowledge as curiosity in myth. Even stronger, gods of knowledge
are sinister and dangerous, such as Thoth, Hermes or Loki,
or as in monotheistic Judaism, where the serpent plays this
role. Only the Greeks accepted a second order goddess, Pal-
las, to preside over knowledge in our sense; but even she
could only be accepted in Athens by providing the olive, i.e.
as technologist.

FIGURE 1

Picture of the Model

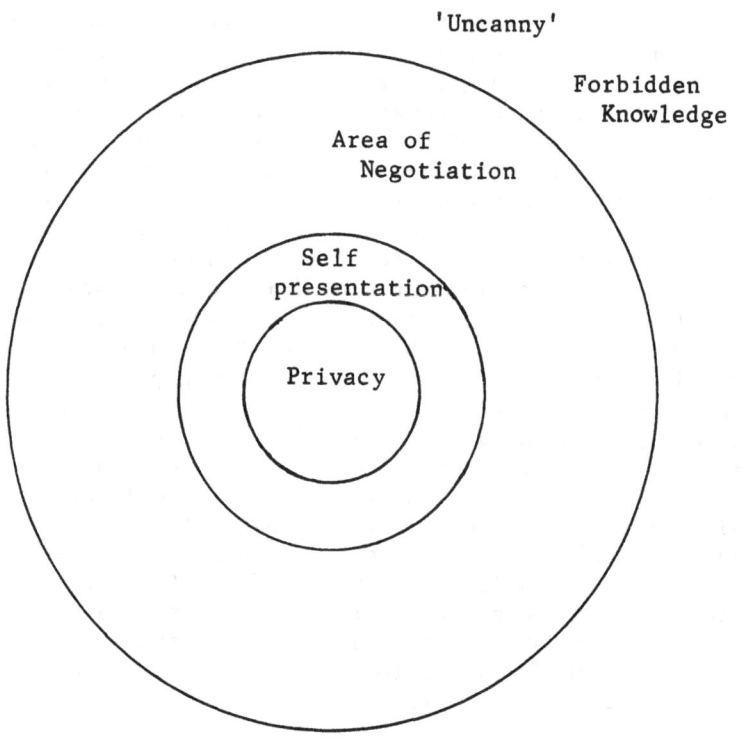

Permitted and Forbidden Knowledge

Let us try now to construct a general model of the relative places of knowledge and blissful ignorance in human consciousness. From this point of view we can visualize topics of knowledge as different regions surrounding each person. Some regions are dangerous to know; some are unknown but can be explored. And some are well-known and even should be known. We can visualize a landscape representing this point of view in a series of concentric circles. There is an area around each person which is considered to be inviolate and inviolable. Here unauthorized knowledge becomes a personal threat. Far out from each person is another region, so remote that it may be dangerous for us to know about. Those are the areas of esoteric knowledge, of knowledge known only to the initiated, which is dangerous for most mortals to know. Between them is the area of knowledge which is possible and can be negotiated, between oneself, other people and society. Partly it includes some knowledge about ourselves which we want to protect, which we reveal only to our intimates. One might even measure intimacy by how much we want others to know about us. Again, some socially privileged individuals might want to know and might have a right to know something about us which we do not want to tell other people. These might be physicians, policemen or census-takers, for example. Around this we can draw an area representing the way we want to present ourselves, what we don't mind that other people know and in fact may want other people to know. This part of the map, the inner circle, is the region where knowledge competes with privacy. But there are also limits to knowledge at the opposite end of privacy, as it were; there are facts which it is judged should stay beyond everyone's reach. This is the area of esoteric knowledge, known only to initiates. Between those limits there is knowledge which should be available to everyone and which everyone should avail himself of the opportunity to obtain (Figure 1).

One feature of a scientific culture is the denial of the separation of the most outward region: the profession that there is no limit to knowledge. Ostensibly, our society does not believe in esoteric knowledge, i.e. facts which should be hidden because of their dangerous and contaminating nature. We may question, however, whether this denial in fact represents the feelings of most members of the society, or-- deep down-- even those of the scientists themselves.

Corresponding to the two boundaries of knowledge, personal privacy and the esoteric realm, we have different types of social control of knowledge depending on personal or social feelings about the propriety of knowledge. Social concern has been expressed, on the one hand, for the protection of the individual against invasion of privacy, the right of the individual to confine information about himself to a chosen few, to assume a reasonable representation of the self. However, even here there may be some social needs which may overrule the individual predilection, such as possession of an infectious disease, or data needed for policy-making. In the conditions where there is a danger to privacy we find restriction of the need to know which may be limited to either a few individuals or categories. We find also the opposite social control which forces everyone to have certain knowledge-- compulsory schooling, compulsory literacy, compulsory civics courses are cases of this kind. Finally, there are social controls which prevent some esoteric knowledge from being known or at least restrict this knowledge to a few individuals and establish controls which should be exerted over those individuals. Although the scientific community does not recognize any limitations on esoteric knowledge, we have analogous conditions in the case of military secrets, whose knowledge may endanger the bearer.

The Cost of Transgressing Boundaries

Societies differ as to the kind of information which is either banned for different reasons, permitted, or required knowledge. As we have said before, our own society is unique in the high value placed on science and on knowledge for its own sake, and therefore claims not to set any limits against the regions of forbidden knowledge. In other societies it is more freely recognized that those limits exist and have always existed. Fear of magic has made people prevent others from having knowledge about themselves. We still feel that knowledge about other people could imply control of them. There are other secrets in a society, preserved as mysteries to which only the initiated may have access; examples of this are the name of God, or the Homeric "drug potent against pain and quarrel", or in our time, the secrets of recombinant DNA or atom-breeding reactors. Everett Hughes (4) has even defined professionals as those people who have access to dangerous secrets, the possession of which gives them the ability to do great harm to the society. Therefore professionals are subjected to strong types of control by the society as well as by their own colleagues and are regarded with a mixture of awe and suspicion.

We may well ask how far these attitudes persist. In our society, many people protest against the conspiracy of science and technology. Many people resent the intrusion of the scientist into private affairs or fear that scientists may go too far-- whatever that may mean in any particular instance.

Even if people do not invoke fear of magic and awe of the unknown, are the reasons which they do invoke for limitation of
knowledge, such as national security, interference with nature or the ecology or the danger to human autonomy, so different in emotional connotations? Each of the various reasons evoked for limiting scientific enterprise may vary according to the preferences of the speaker and the audience to
which his argument is addressed.

We can propose, then, that there are limits built into
human cognition and that there is a cost attached to going
beyond them or of trying to deny them. And thus we have seen
even in our own society pure knowledge and basic science fal-
ling into low esteem, having degenerated in the popular imagination into associations with either idle curiosity or the
stereotype of the mad scientist. This is easily indicated
by the fact that financial support in Congress for basic science programs has to be bootlegged through projects promising immediate material payoff.

The existence of stringent social controls is not as
surprising as it may look to the dedicated scientific worker.
The value of knowledge and of truth, which scientists take
for granted, may not only rank low in comparison to other
values but is seen as a danger in many contexts. None of
the Ten Commandments enjoins lying. History could show more
tradition on the side of those who would want to suppress
scientific knowledge, except if its value is shown for other,
more practical reasons, than those in favor of the unrestricted search for knowledge.

Let us summarize the discussion up to this point. We
have seen two aspects of the deep-seated emotional fear which
opposes the extension of knowledge. One is in the individual, in the fear of the invasion of the self and of privacy and the other is the fear of the outer reaches of knowledge, of a secret kind of region where knowledge might invade the privacy of something greater. There are different
kinds of problems associated with each of the limits of legitimate knowledge, and different social conditions in which
they become important. In our society the balance is toward openness and toward knowledge and we are more hospitable
to the expansion of knowledge than most cultures. We have
our own problems with the limitations of knowledge, however.
They derive in part from the nature of our society, from social conditions and partly from the predilections of individuals.

Social Conditions of Boundaries

Social restrictions on the pursuit of science are not

TABLE 1

A Classification of Societies
(After Douglas)

GROUP

		High	Low
	High	Tribal Society	Insulation Passivity
GRID			
	Low	Small Group	Rationalized Individualistic Society

just the weapons of enemies of progress. As we have shown, limitations on knowledge are inherent in the structure of human personality and society. Thus we shall relate the problems of limits of knowledge to the general role of boundaries in different societies. Social anthropologists and structural theorists have developed classifications of the different boundaries in society which will be useful at this point; we can then discuss some examples of reluctance to step over these boundaries.

Knowledge can be seen as a willingness to incorporate the objective world into oneself and also the willingnesss to accept being an object of apperception for other people. It is a willingness to transcend boundaries. A social definition of these boundaries and their strengths can determine the conditions under which societies will accept and reject knowledge. In defining a set of possible boundaries, a system of conditions devised by Mary Douglas and Basil Bernstein has been useful in making a classification (3,2).

This classification is based on individualism as well as on group behavior. The dimension of individuation is called 'grid,' the strength of "rules to which individuals are subject in the course of their interaction" (3, p. 8). In high grid conditions the individual person is classified by ascribed features, leaving little room for personal choice of behavior; in low grid conditions individual initiative is maximized, and expressed through such mechanisms as autonomy, control and competition. 'Group' is defined by the investment of time and energy of the individual members and agreement among the group members as to the boundaries of the group— who is in and who is out. High group conditions mean complete involvement in one group for all activities and definite rejection of all people outside its well-defined boundaries; low group is a condition in which "a person finds himself the center of an environment of his own making which has no recognizable boundaries" (3, p. 16). Both group and grid define possible social environments (Table 1).

These two dimensions of the social environment have consequences for many social patterns, including the control of knowledge. The separation of the in-group from the rest of the world will also include rejection of all knowledge which is not directly a part of the culture of the group; thus the boundary against forbidden knowledge will be strong, and the domain of this prohibition will be large. Social situations with high group dimensions will have a minimum of knowledge about the world and strong defenses against extending it.

Grid conditions will be relevant to the boundaries

around the person. But they are not simply analogous to the
relations of the outer boundary to group dimensions. High
grid conditions imply the importance of self-presentation, as
interaction is governed completely by the assignment to social
position which depends on the social self. Low grid conditions, however, do not imply an open self. Interaction becomes intensely competitive and private knowledge is increasingly valued as a resource.

We may suspect from the definition of group and grid that
we will find that these conditions are correlated. Societies
which distinguish themselves from the rest of the world also
have many distinctions within them, and societies which don't
do so look at their own members mainly for what they can do,
not for what they are. Frequently, we find stable permanent
arrangements as societies where either group and grid are low
or both are high. However, having boundaries, as well as
transcending them, has its cost. There are always some pressures on and by the individual to vary the socially determined boundary system; within each type of society there may
be a reaction which changes the type of organization structure. We can look at the high-high and the low-low societies
and see how they are transformed individually or collectively.

Societies in which both group and grid are high are what
are called tribal or traditional societies. Here everyone
has a distinct position and the boundaries are well-kept.
The tribe and the rest of the world are rigidly separated.
Each status is clearly defined; transitions from one status
to another are marked by definite changes and ceremonies;
contacts between persons in different statuses are well-reg-
ulated. There are definite joking, aggressive or emotional
relationships between statuses.

The low-low society corresponds to the extreme modern
society. Here the society is open to other cultures and
trades knowledge and traditions. Within the society statuses
are defined only according to the functions a person can perform and do not impose any barriers irrelevant to the performance. Professionalism and the impersonality of modern society are consequences of this low group and low grid condition. For instance, the provider of medical care is a person trained in a certain way, independent of sex, race or
family background. There is no special sacred knowledge
which is peculiar to a particular status and must be preserved. We have Popular Science, Popular Mechanics, Psychology Today instead of the arcane volumes of the alchemists
and magi. Knowledge becomes a commodity which has lost its
threatening and sacred character. On the other hand, we increasingly have laws and regulations protecting individual

privacy. One of the recurring and perhaps unavoidable conflicts in our society is between open knowledge as a public
good and privacy as an individual asset.

Like any pure concept these extreme conditions are seldom completely realized. Impenetrable barriers as well as
complete openness are not found in human society; being completely enclosed or being completely open seems to put too
much strain on the individual. Deviations occur and reactions set in. We can find conditions where either grid or
group is more emphasized, as in transitions between tribal
and modern society or as reactions to them. Two additional
types can be distinguished. If the grid condition, the interpersonal relations alone, gets emphasized, we have societies which are open to an outsider but extremely careful and
strict in the behavior of the individual members toward each
other. We find a romantic revolutionary situation which
wants to create a new world for everybody, where the behavior
of the members toward each other becomes of extreme importance, and is definitely controlled. Thus the social self--
self-presentation--becomes crucially important and well-
defined. There is corresponding pressure on social rules to
make self-presentation identical with the private self, i.e.,
to make the private self open to inspection. Authenticity,
public self-analysis and communal living become accepted life
styles. On the other extreme, we have the societies which
over-emphasize the group conditions. Those societies would
emphasize duty toward society, suspicion toward outsiders,
use magic to protect their personal boundaries, and distrust
any new ideas coming in or their own knowledge going out.
These societies accept structure; the role of the individual
is over-emphasized. These different conditions will characterize deviant phenomena in the tribal as well as in the modern society.

Modern Society and its Enemies

Thus, in general, the ideal of the scientific society is
one which has low grid, low group, the ideal type of modern
society. However, the reactions against this society of open
knowledge come from different directions, which we have sche-
matized as desires for different boundaries, group and grid.
Threats to knowledge and to the freedom of science do not
come from one side or one set of villains. Neither the left
nor the right, neither the military establishment nor the
free populace has any monopoly on the wish to restrict the
freedom of collection and dissemination of knowledge. These
restrictions, the new boundaries, testify to the precarious-
ness of the values of knowledge in general in the human
scheme of things. We can take as a primary hypothesis a human need to establish boundaries, certain limits, and thus

expansion in one direction may be compensated by limiting another. In the public use of language we have eliminated censorship of scatological or blasphemous terms; practically the same people who have fought for this expansion of linguistic boundaries are quite willing and eager to impose other censorship under the name of eliminating sexism, racism or any other isms they can think of. In the same way the whole fabric of the control of science seems to be interconnected, winning at some points and losing at others. We cannot talk about a pure, rational model of science and what would be best for the growth of science in the abstract. We must in addition look at individual needs and at the restrictions individuals put on knowledge for different reasons. We shall therefore proceed from discussions of whole types of societies to individual data on the acceptability of the acquisition of knowledge. A study of this kind would have to include perspectives on knowledge of different kinds, on preservation of privacy and on the outer limits of science, their interrelations and their conditions. This is a large program which cannot be completed quickly. I shall present here an attempt to start with a part of this question and analyze the conditions under which people accept certain invasions of privacy for the dissemination of knowledge.

A Study of Privacy and Interviewing

One of the places where the question of privacy interferes with scientific research is the collection of personal data through interviews. Interviews are conversations with a purpose and we can assume that, depending on the purpose, different degrees of invasion of privacy will be acceptable. For this study, four kinds of purposes for conversations were selected. One was pure conversation, conversation between friends in a social situation. This in a way is the control condition. Other conditions were the personnel interview or the medical background interview as conversations with definite ulterior purposes. Finally the interview could be conducted for purely scientific or social science purposes, as in a survey interview.

We constructed an interview which could have been applicable to any of those four conditions, consisting of forty questions and answers, starting with "Where are you living now?"-- "I live on West Main Street"-- and ending with the more restrictive question "Have you ever taken any illegal drugs, narcotics, or prescriptions that weren't yours" with an answer "I tried pot once or twice," etc. These questions could be grouped into ten topics--general background, drinking and smoking, anxiety, working life, money, marriage and sex, social appearance, communications problems, general

health and drug use. Subjects were asked to read the whole
interview, to cross out statements about things that they
would not want another person to know, to underline the
statements about things they would want another person to
know and to leave neutral the rest.

The sample population consisted of three groups that
could be considered to be experts in each field of interviewing, namely, personnel managers, physicians and survey interviewers. The first group was selected from a compendium of
the largest American corporations which gives a list of their
personnel managers. The physicians were selected from a directory of physicians using general practitioners and psychiatrists as those people most conscious of interviewing. Survey interviewers were approached through the help of three
major survey organizations.

Each of these subject groups was asked in one version
whether they would feel that they shouldn't ask a particular
question or whether they would want to ask a particular question. This was done in the context of their own profession--
for instance, with physicians on a medical interview. Where
the interview was presented in the other three contexts, they
were told to identify with the respondents and were asked how
they would feel as subjects of the interview. Each person
filled out only one version of the questionnaire. Thus each
physician, personnel manager or interviewer could get one of
four versions of the questionnaire. For instance, physicians
could get a questionnaire where the interview was presented
as a medical interview; there they would be asked how they
would feel if they were conducting the interview. They could
also be given the version in which they were told that this
was a personnel interview, a survey interview or meeting an
old, long-lost friend. In these cases, they were treated as
the interviewee and were supposed to identify themselves with
the interviewee. It is obvious of course that these interviewees were not a good sample of the possible interviewees;
on the other hand, they give a good control for the role of
interviewer and interviewee. We have thus a three by four
design-- three kinds of subjects and four possible conditions
for each subject. The questionnaires were sent out by mail.
Although the response rate was quite low, we are not too concerned here about representativeness to a population. For
each type of subject-- personnel manager, interviewer and
physician-- we thus have indications of what kind of information each considered to be appropriate to his own field and
that which seemed relevant to other fields. In analyzing the
data we find that in practically all of the topics we have
strong and statistically significant differences in both directions, by subject and by situation (Table 2).

TABLE 2

Acceptability of Questions

	Deviation from Mean						Significance			
	By Sample			By Situation						
	M.D.	Personnel Manager	Interviewer	Medical Personnel	Survey	Friend	F	p	F	p
Background	-.71	3.73	-1.92	-2.25	3.14	-.80	Sample F = 22.40	<.001	Question F = 6.13	<.001
Drinking and Smoking	-1.40	1.97	-.71	-3.31	2.42	-.24	Sample F = 15.72	<.001	Question F = 31.51	<.001
Anxiety	-.61	.81	-.28	-2.02	.66	.02	Sample F = 8.26	<.001	Question F = 25.37	<.001
Working Life	.49	-.29	.02	-.05	-.78	-.07	Sample F = .66	.52	Question F = 10.19	<.001
Money	.93	.47	-.53	.55	-.73	-1.08	Sample F = 4.54	<.01	Question F = 16.84	<.001
Marriage/Sex	-.77	1.35	-.55	-1.60	1.59	-.72	Sample F = 7.45	.001	Question F = 12.26	<.001
Social Appearance	.00	3.16	-1.79	-1.83	3.60	-1.64	Sample F = 19.18	<.001	Question F = 13.56	<.001
Communication	.41	1.63	-1.04	.04	1.45	-1.32	Sample F = 12.80	<.001	Question F = 5.98	.001
Health	-.27	.97	-.47	-1.69	.92	-.26	Sample F = 6.15	.002	Question F = 12.43	<.001
Drug Use	-.44	.99	-.44	1.82	1.13	-.45	Sample F = 10.37	<.001	Question F = 31.63	<.001

Who Can Ask Questions

Let us look first at the situations. On seven of the ten topics all respondents considered the physicians' inquiries to be the most legitimate. On questions of work history and income, the personnel interview was considered to be most legitimate and on questions of communication difficulties, such as "Do you express worries to others?", the survey interviewer was considered to be the most legitimate questioner. Within very narrow ranges of relevance to interviewees' jobs, personnel managers were most accepted, and in the discussion of communication itself, interviewers are the most relevant questioners. Otherwise everyone accepts the physician as a legitimate questioner. In general, the search for knowledge alone is not a sufficient reason for the invasion of privacy. The boundary is most relaxed when the respondent can get something out of the experience. However, knowledge acquisition by the survey interviewer is not definitely rejected. The survey interview ranks decidedly up in second place in the scoring of acceptability. Thus the rejection of questions, which was expressed by lining out answers, was as pronounced in a circumscribed quasi-legal situation as in the purely social situation. Respondents did distinguish between the aims of science and idle curiosity. We might say, if we consider the position of the interviewer in this situation as the surrogate of the scientist, that the search for knowledge is not particularly approved but rather accepted as a fact of life and sharply distinguished from the avid curiosity which might be shown by a friend. We can also distinguish acceptability by the type of respondent involved. We find here a slightly different picture. The respondents are all professionals in one field of interviewing and we might look at the answers here as differences in professionalization on the general legitimacy of obtaining personal knowledge. This is confirmed by the fact that with one exception the differences between the different groups are highly statistically significant. The survey interviewer turns out to be most accepting of asking personal questions. In five fields the interviewer has the highest acceptance score, the physician has it in four and only in one field, in the question about work, does the personnel interviewer score highest, and this is a non-significant difference. That is, if we regard the situation in which the question is asked, the interviewer will be in general the most accepting of asking questions, a little more than the physician. This is different from the results we got on situational variables. We may interpret the findings to mean that the interviewers are socialized in general to place a higher value on knowledge. In regard to situations, however, the medical situation is the one where knowledge is most easily exchanged about a person.

We can summarize these findings on the two factors by looking at the difference when a person identifies with the questioner and when he identifies with the respondent. Each of the three respondent groups had one condition which corresponded to their own profession and in this they were identifying with the professional and not the respondent. Comparing the respondent groups according to this difference we find a great contrast between the interviewers and the personnel managers, with the physicians in between. On eight out of the ten topics the personnel managers felt more cautious as interviewers about the respondents' privacy than the respondents themselves were. The situation was reversed completely, however, with the interviewer. The interviewer was much more willing to invade privacy in all ten topics than the respondents were willing to give information to the interviewer. And here we see the contrast-- the interviewer was more socialized to the right for knowledge than the respondents were to accepting these claims of pure science.

The physician was in between. In eight of the ten situations the physician required more openness than he was likely to receive. However, we must remember that the medical situation was by general agreement the area in which invasion of privacy was most likely to be tolerated, while in the survey interview there is a great difference between the interviewers' claims and the respondents' willingness to provide information.

There may be, of course, other differences between the groups than professionalization into the value of knowledge. Physicians, for instance, are the oldest and best educated group; personnel managers are the youngest. The interviewers are the least educated; they were also most likely to be women. On self-rating of political views from "very liberal" to "very conservative" physicians were clearly conservative; the other two groups were similar, with interviewers slightly more liberal. The study also contained a scale of general attitudes toward individual boundaries with such questions as "Should people be more open about themselves?" or a statement such as "It is important to keep one's distance from other people," which demanded a positive or negative response.

This scale relates to all topics discussed in the interview transcript; favorable attitudes toward strong boundaries correlate to rejection of questions about oneself. The scale represents a general rejection of invasion of privacy for any reason, which can, however, be modified by the specific situation, as the correlation is by no means perfect.

Two background factors related strongly to the scale:
one was the sample source--physicians, personnel managers or
interviewers; the other was political self-identification,
from very liberal to very conservative. Physicians were most
distinguished from the other two groups by resisting invasion
of privacy; conservatives also put a high value on privacy.
However, the two effects were not independent: the influence
of political views on need for privacy occurred almost exclusively among physicians; or, looking at the data in another
way, the difference between the three sample groups occurred
principally among conservatives. (Statistically speaking,
the interaction between the political view and the sample
source was significant.) To add to the complication of these
data we must remember that the medical situation was the one
in which-- by general agreement-- the most invasion of privacy was allowed.

These data indicate that a study of the restriction of
scientific research can be fruitfully approached through general studies of boundaries, in individual cognition and social rules, of conditions under which they can be transgressed, the cost of transgressing them, and the values acceptable for their transgression and violation. This general sociological framework can give a secure base for the understanding of similar cases of social control and restriction
of science.

References

1. G. Allport and P.E. Vernon, _A Study of Values: Manual of Directions_ (Houghton-Mifflin, Boston, 1931).

2. B. Bernstein, _Class, Codes and Control_ (Routledge, Kegan Paul, London, 1971).

3. M. Douglas, _Cultural Bias_ (Royal Anthropological Institute of Great Britain and Ireland, London, 1978).

4. E. C. Hughes, "The Study of Occupation" in R. K. Merton, L. Broom, L.S. Cottrell, Eds., _Sociology Today_ (Basic Books, New York, 1959), pp. 442-458.

5. C. Morris, _Varieties of Human Value_ (University of Chicago Press, Chicago, 1956).

6. M. Rokeach, _The Nature of Human Values_ (Free Press, New York, 1973).

7. E. Spranger, _Types of Men_ (Max Niemeyer, Halle, 1928).

Part II

Regulation of Scientific Inquiry in the Context of Recombinant DNA Research

8

Recombinant DNA Legislation

Adlai Stevenson

One of the more important issues on the congressional
science and technology agenda has been consideration of
legislation to regulate the conduct of scientific research
involving DNA recombinant molecules.

Research involving DNA recombinant molecules could well
unlock the mysteries of heredity and life itself. Industrial
technologies based on recombinant DNA techniques could bring
previously unimagined benefits to mankind in such diverse
fields as medicine, agriculture, and energy. Recognition of
these potential benefits has been accompanied by an equally
strong fear of the possibility of creating new forms of life
that could threaten humanity. The heart of our legislative
task has been to devise a way to reap the benefits of
recombinant DNA research while protecting mankind against
some biological catastrophe.

As we consider our present course of action, three considerations must be kept in perspective.

Initial Concerns of Scientists

First, we should bear in mind how this issue first came
to public attention. More specifically, what were the
initial concerns of the scientific community? What dangers
did these scientists seek to avoid? What actions did they
recommend and were these actions taken?

Answers to these questions provide an important backdrop
to whatever decisions Congress makes regarding the regulation
of recombinant DNA research. It is clear that scientists
were the ones who first spoke out. It is also clear that
they were primarily concerned about potential, not verified,
risks associated with recombinant DNA molecule experimentation. For instance, Dr. Maxine Singer of the National

Institutes of Health, in the statement prepared at the 1973
Gordon Conference on Nucleic Acids, said:

> Certain such hybrid models may prove hazardous to
> laboratory workers and to the public. Although
> no hazard has yet been established prudence suggests
> the potential hazard be seriously considered.

The Gordon Conference statement further recommended to
the President of the National Academy of Sciences and to the
President of the National Institute of Medicine that--

> The Academies establish a study committee to con-
> sider this problem and to recommend specific actions
> or guidelines, should that seem appropriate.

Such a committee was created and in July 1974 its
report was published in Science magazine. The report noted:

> Our concern is based on judgments of potential
> rather than demonstrated risk since there are few
> available experimental data on the hazards of
> such DNA molecules.

The committee went on to urge all scientists working
with recombinant DNA molecules to avoid such experiments
until further attempts had been made to evaluate the actual
hazards that might be involved.

These early initiatives by the scientific community
reflected a well-defined sense of public responsibility.
The plea of these scientists was clear and straightforward:
exercise great caution in this new area of experimentation
until the potential risks can be evaluated further. There
was no claim, at any time, that such experiments should be
banned permanently; nor was there any prejudgment as to the
level of risk that inevitably would be involved in recombinant DNA research.

These initial expressions of caution were coupled with
recommendations to devise experimental procedures that would
protect the health and safety of recombinant DNA researchers,
the general public, and the environment. Again, the scientific community responded. In February 1975, The Asilomar
Conference, sponsored by the National Academy of Sciences,
proposed physical and biological containment standards for
recombinant DNA research. The report of the Asilomar Conference noted:

The evaluation of potential biohazards had
proved to be extremely difficult. It is this
ignorance that has compelled us to conclude that
it would be wise to exercise considerable   caution in performing this research . . . the standards of protection should be greater at the
beginning and modified as improvements in the
methodology occur and assessments of the risks
change . . . future research and experience may
show that many of the potential biohazards are
less serious and/or less probable than we now
suspect . . . This document represents our first
assessment of the potential biohazards . . .
little is known (about survival of laboratory
strains of bacteria, enhancement effects,
depression of survival, etc.) . . . answers
are essential to assessment of biohazards.

The dominant theme of caution, the need for further
experimentation, and an overriding concern for safety were
again very clear. But the possibility of modifying these
standards to reflect new research findings and experience
was explicitly acknowledged.

Accumulation of Scientific Evidence

The second consideration to keep in perspective is the
scientific evidence that has accumulated since 1973. In the
process of formulating the research guidelines published by
NIH in July 1976, the NIH DNA Advisory Committee conducted a
series of technical meetings throughout 1975 and 1976. These
sessions have continued in 1977 to consider possible revisions of the guidelines. During these deliberations, careful attention was paid to learning more about the surviv-
ability of special "biological containment" systems, such as
E. coli, as well as to refining the proper levels of physical
containment for different types of recombinant DNA experiments.

As these efforts continued, it became increasingly clear
that many of the hazards associated with recombinant DNA
research were less severe than initially feared. In time,
scientists began to report these changes, revising downward
their perception of risk and suggesting the feasibility of
adjusting the NIH guidelines accordingly. Moreover, scientists were also discovering that many of the so-called novel
DNA recombinations were nothing more than laboratory-con-
trolled duplications of events which could occur in nature.
These later discoveries raised questions about the proper
definition of recombinant DNA that should be used as a

basis for any Federal regulatory effort.

These new perceptions were summarized by Dr. Roy Curtiss III, University of Alabama, in a letter of April 12, 1971, to Dr. Donald Frederickson, Director of NIH. Dr. Curtiss described his experiments which had been conducted to determine whether a DNA recombinant molecule in an E.-coli host could be transferred to other microorganisms if accidentally released from the laboratory environment. He also reported similar experiments conducted by other DNA researchers. On the basis of this new evidence, Dr. Curtiss concluded that he could no longer perceive any danger to human beings or the environment, assuming that normal experimental safeguards were followed. Curtiss wrote:

> The arrival at this conclusion has been somewhat painful and with reluctance since it is contrary to my past "feelings" about the biohazards of recombinant DNA research . . . In no instance-- in this and other experiments--have I found evidence that the necessary genetic information could be cloned in one step, would permit E.-coli K-12 to colonize the intestinal tract, and to lead to the production of the product(s) in the intestinal environment that would be harmful to the mammalian host.

The substance of these findings and the rapidity with which they were being made raised new doubts about the various regulatory approaches being considered by Congress. For example, many of the initial assumptions concerning the scope of the regulations were open to question. Some scientists viewed the physical and biological containment standards adopted by NIH in its July 1976 guidelines as excessively stringent. Revisions of the guidelines were contemplated by NIH. Concern was also expressed about legislative approaches that would subject recombinant DNA researchers to the redtape and complicated procedures of a Washington-based regulatory commission. University-based scientists were particularly concerned about the threat of criminal liability, as provided for in some legislation, on the research process. Although it was important that regulatory controls be extended to privately supported research, these controls should be administered in a way that did not place an unequal burden on smaller private research firms. Other issues, such as the protection of proprietary information and scientific discoveries, remained a source of contention.

In these circumstances, Congress has every reason to

proceed cautiously in writing permanent legislation. There is a clear need to take account of the rapidly changing perceptions of the likely risks of recombinant DNA research. This view is supported by a new substantive evaluation of the problem being conducted by the Carter administration.

My hope is simply that Congress can be as responsive and as concerned with scientific evidence as were the scientists who initially spoke out about the potential risks of recombinant DNA research.

Much of this analysis has been drawn from an excellent summary of actions affecting the perception of risks in DNA recombinant molecule research prepared by the Congressional Research Service.

Their summary of actions affecting the perception of risk in DNA recombinant molecule research is as follows:

Developments in molecular biology which were discussed during the June 11-15, 1973 Gordon Conference on Nucleic Acids (New Hampton, New Hampshire) led to the issuance of a statement by some of the conference participants concerning DNA recombinant research. In this statement, signed by Dr. Maxine Singer of the National Institutes of Health and Dieter Soll of Yale University, on behalf of a number of scientists at this conference, the technique of DNA recombinant research was identified and a comment included that:

"Certain such hybrid models may prove hazardous to laboratory workers and to the public. Although no hazard has yet been established, prudence suggests the potential hazard be seriously considered."

This concern was communicated in this Gordon conference letter to the President of the National Academy of Sciences and to the President of the National Institute of Medicine. The suggestion was made that:

" . . . the Academies establish a study committee to consider this problem and to recomment specific actions or guidelines, should that seem appropriate."

A committee (known as the "Berg Committee" since it was chaired by Paul Berg of Stanford University), appointed by the National Academy of Sciences to investigate the concerns expressed in the "Gordon Conference" letter, conducted a preliminary study of the problem and made its report to the Academy and the public. In their report, (published in Science in July 1974) the committee noted, among other ponts, that:

" . . . our concern is based on judgments of potential rather than demonstrated risk since there are few available experimental data on the hazards of such DNA molecules and (ii) that adherence to our major recommendations will entail postponement or possibly abandonment of certain types of scientifically worthwhile experiments . . ."

"Nonetheless, our concern for the possible unfortunate consequences of indiscriminate application of these techniques (DNA recombinant research) motivates us to urge all scientists working in this area to join us in agreeing not to initiate experiment . . . (of certain types) . . . until attempts have been made to evaluate the hazards and some resolution of the outstanding questions has been achieved . . ."

Then, in February 1975, an International Conference on Recombinant DNA molecules was held at the Asilomar conference center (California). This conference was sponsored by the Assembly of Life Sciences, National Research Council, National Academy of Sciences and a summary report of the conference was published in the Proceedings of the National Academy of Science in June 1975. The meeting was limited to invited participants including a select few lawyers and press representatives. In their summary report, the committee provided recommendations with regard to the physical and biological containment which should be utilized to insure safety. Categories of containment were described which are now further refined and identified with the P1-P4 codes established in the NIH guidelines and the concept of biological containment, now known as Ek-1, Ek-2, etc. in the NIH Guidelines was anticipated by the committee. In the introduction to their recommendations, the conference committee indicated that the meeting:

" . . . was organized to review scientific progress in research on recombinant DNA and to discuss appropriate ways to deal with the potential biohazards of this work."

The committee report also noted that:

" . . . the evaluation of potential biohazards has proved to be extremely difficult. It is this ignorance that has compelled us to conclude that it would be wise to exercise considerable caution in performing this research . . . the standards of protection should be greater at the beginning and modified as improvements in the methodology occur and assessments of the risks change . . . future research and experience may show that many of the potential biohazards are less . . . represents our first assessment of the potential biohazards . . . little is known (about survival of laboratory strains of bacteria, enhancement effects, depression of survival, etc.) . . . answers are essential to assessment of biohazard."

Following the Asilomar conference, the NIH DNA Advisory Committee began its series of intensely technical meetings to construct detailed guidelines. This committee was chartered by the Director of the National Institute of Health in October 1974. In their charter, it was noted that:

" . . . the use of this technology, (DNA recombinant molecule research) has various possible hazards because new types of organisms, some potentially pathogenic, can be introduced into the environment if there are no effective controls . . . The goal of the Committee is to . . . recommend programs of research to assess the possibility of spread of specific DNA recombinant molecules . . . and to recommend guidelines on the basis of the research results."

At the same time, as information became available, there was an obvious increase in awareness, with emphasis upon the variants of Escherichia coli which were being constructed as cloning systems, that many of the hazards originally postulated for such systems were highly improbable. This emphasis evolved in association with the conduct of specific types of experiments not prohibited by the guidelines. From these many exchanges among professionals during the development of the current NIH guidelines, as well as in the work directed toward the revision and updating of the guidelines, changes in the perception of risk became evident.

It also became obvious that the definitions of the types of DNA recombinant research covered by the NIH guidelines required more specific language since many of the experiments being conducted were not really "novel" recombinations but were laboratory controlled duplication (although accelerated) of events which were possible and probably occurred in nature. Some investigators, including Stanley Cohen at Stanford, became more certain as they continued their evaluations of the biochemistry of DNA recombination that there was indeed a high degree of recombination going on in nature by the same biochemical mechanisms as the experiments being conducted in the laboratory. This concept of non-novel recombination and the increasing certainty about the virtual absence of risk when using E. coli variants and non-prohibited types of experiments became further strengthened by other evaluations.

An analytical paper which had a major impact within the scientific community was a letter submitted to the Director of NIH by Dr. Roy Curtiss III of the University of Alabama. In his letter of April 12, 1977, Dr. Curtiss described the many experiments conducted in his laboratory to evaluate the potential risk if a DNA recombinant molecule in an E. coli variant was accidentally released or whether the recombined molecule could be transferred to another microorganisms as a result of an accident. In his letter, Dr. Curtiss also summarized a number of other risk assessment experiments which had been conducted around the country and pointed out that these data consistently supported his own opinion that he could no longer perceive any danger whatsoever to any human being with the exception of the risk to a careless laboratory worker, as always exists in any work with microorganisms.

The initial view that experiments with DNA
recombinant molecules might pose some unknown
risk was a reaction primarily by molecular biologists and bio-chemists who were not familiar
with the detailed epidemiology or ecology of the
"model" systems with which they were working. As
a part of the continuing evaluation of the risk,
an effort has been made in the past several years
to involve those scientists who do have the
experience in working with infectious organisms
to estimate the dangers which might be produced
by research with recombinant DNA molecules.
These groups have been publishing their views
very recently, and their views also have had an
impact on the DNA recombinant research scientists'
perception of risk. For example, at their annual
meeting in New Orleans in May 1977, the Council
of the American Society for Microbiology adopted
a resolution which included a recommendation that
experiments requiring proposed P-1 containment
should be exempt from the regulations for DNA
recombinant research. In a letter to <u>Science</u>
from the President of ASM (Harlyn O. Halvorson,
Brandeis University) it was noted that:

" . . . there has been surprisingly little
input from individuals accustomed to dealing with
hazardous microorganisms. . . . There is at
present no demonstrable evidence that microorganisms containing recombinant DNA molecules
are hazardous."

Following the 1977 Gordon conference, an
open letter to Congress was submitted from a
group of scientists which was presented as a
consensus of opinion. This letter, published
in <u>Science</u> on July 15, 1977, was signed by 137
members of the conference (86 percent of the
members of the meeting). Comments included in
the letter indicated that:

"We are concerned that the benefits of
recombinant DNA research will be denied to
society by unnecessarily restrictive legislation
 . . . We feel that much of the stimulus for
this legislative activity (legislation to regulate DNA recombinant research) derives from
exaggerations of the hypothetical hazards of
recombinant DNA research that go far beyond any
reasoned assessment . . . the experience of the

last four years has not given any indication of actual hazard. . . ."

Still later (July 14, 1977), Sherwood L. Gorbach (New England Medical Center Hospital and Professor of Medicine and Microbiology, Tufts University School of Medicine) provided a letter to the Director of the National Institutes of Health in which he reported what he called a "consensus of opinion" following a Workshop of Studies for Assessment of Potential Risks Associated with Recombinant DNA Experimentation (now referred to as the Gorbach Falmouth Report.) The Workshop (held at Falmouth, Massachusetts) was sponsored as a part of the NIH program to secure risk assessment data for input to the restructuring of the DNA recombinant molecule research guidelines. His concluding statement was:

"It was the consensus of the group that the possibility of transfer of a non-mobilizable plasmid from E. coli K12 to a wild-type strain in vivo is extremely unlikely. However, it was recommended that animal and human feeding experiments should be performed, to confirm the previous observations in man and to provide further assurance that in vivo transfer of these plasmids should not occur."

In evaluating these trends in risk assessment, it should be kept in mind that primary attention has been given to the E. coli experiments and not to other host cell systems. Since the vast majority of DNA research molecule experiments are with E. coli variants, the experimenters actually are saying that these DNA recombinant experiments pose little, if any, risk and therefore there is no need for strict regulation. This is not to say, however, that these investigators are suggesting at this time that all of the prohibited experiments of the NIH guidelines now should be permitted although, in some instances, evidence is being cited, as with the transfer of toxins through E. coli variants, that there may be little risk even with certain types of these experiments. The efforts at reconstructing definitions of the types of DNA recombinant research to be covered by the NIH guidelines which have been occurring within the NIH DNA Advisory Committee also would reflect these new perceptions

of risk and exclude those experiments which are not
novel, that is, experiments where it is known that
exchanges occur in nature.

Freedom of Scientific Inquiry

The third consideration to keep in perspective is our
responsibility to safeguard the conduct of scientific
research from unnecessary and unreasonable governmental
intervention.

Freedom of scientific inquiry is rooted in the first
amendment to the U.S. Constitution as a fundamental element
of freedom of speech--Sweezy against New Hampshire; Keyishian
against Board of Regents; Epperson against Arkansas. Of
course, such freedom is not absolute. The Atomic Energy Act
imposes stringent regulatory controls over the scientific
use of radioactive substances. Controls are enforced in the
use of human subjects in scientific experimentation. Dangerous drugs and chemicals must be used according to governmental standards. There is, however, a strong commitment
under the Constitution to the principles of free inquiry.
There is, or at least there should be, a heavy presumption
against governmental interference. This is particularly
true in relation to regulation of recombinant DNA research;
this is the first time that government is proposing to regulate actual research procedures, as distinct from materials
or equipment used in research.

The constitutional protections afforded scientific
research require that governmental regulation be justified
by a compelling need to protect the community's health and
welfare and that governmental regulations be drawn as narrowly as possible to achieve the necessary protection.
Dr. Thomas I. Emerson, Lines professor of law emeritus,
Yale Law School, recently addressed this issue in testimony before the Subcommittee on Scientific Research of the
House Committee on Science and Technology. Professor
Emerson said:

> The degree of risk that would be needed to
> justify governmental regulations could only be
> determined in the light of concrete information
> and presumably the more serious the risk, the
> less degree of certainty that would be demanded.

The recent evidence of the decreased risks associated
with recombinant DNA research using E. coli K-12 as the
host vector requires us to weigh carefully the benefits of
the proposed regulations against their likely impact on the

freedom of scientific inquiry. At this juncture, one can
easily rule out extreme remedies: Few persons would argue
in favor of eliminating all regulatory controls. Likewise,
a total ban on recombinant DNA research, or even highly
stringent regulations, no longer appears either necessary
or reasonable. The most apparent solution would seem to be
the extension of the NIH guidelines, as revised, to cover
both federally supported research and the private sector.
In striking this balance, Congress must take fully into
account the protection which scientific inquiry is afforded
under the Constitution.

Regulation of Recombinant DNA Research 9

An edited version of a statement given by
Philip Handler to the Senate Subcommittee on Science, Technology, and Space of
the Committee on Commerce, Science,
and Transportation*

I guess the difficult problem for the interested lay-
man, by which I mean anyone, scientist, or anyone otherwise
who has not been professionally close to this field
(recombinant DNA research) has been to obtain a sense of
where the bulk of informal opinion lies, to distinguish
statements of fact from science fiction.

In the chorus of protest against conduct of this
research, only a few themes are to be found. Let me
summarize them.

One, the earliest concern was for the possibility that,
in the course of such research, there might be generated a
micro-organism which would escape laboratory control and,
being of a new species, might prove to be dangerous to man,
to domestic animals or to the environment. The possibility
of such pathogenicity has been seriously considered by
responsible and knowledgeable scientists, as well as by
individuals more broadly concerned with protection of the
environment. I shall return to this topic shortly.

Appreciative of the experimental power of this new
laboratory technology but looking further into the future
than working scientists might be willing to venture, some
suggest that this research could be a major step along a
trail that, ultimately, would lead to the capability of
genetic manipulation of man himself. Out of belief in the
technological imperative--"That which can be done will be
done"--and our shared repugnance at the idea of genetic
manipulation of human beings, they argue that it would be
best to prevent the development of this capability by
halting all possibly contributing research at its earliest
stage, viz., research on recombinant DNA. Their position,

*November 2, 8, and 10, 1977, Serial No. 95-52, Washington:
U.S. Government Printing Office, 1978, pp. 4-17.

in short, is that there are some facts that man should not seek to learn.

The third theme--theological metaphysical, mystical, or naturalistic in varying degree--urges that we do not do anything that smacks of tinkering with evolution not cross some imaginary barrier between eukaryotic and prokaryotic cells. The advocates of this view would avoid creation of strains of bacteria with a capability for synthesis of proteins not normal to them--specifically Escherichia coli K-12 the organism adopted for use in these studies.

Thus, whereas the first concern was simple protection of the public health, the latter two seek abridgement of the freedom of scientific inquiry. I shall return, shortly, to the matter of risk; but this biochemist, with apologies to students of constitutional law, would like first to consider briefly the matter of freedom of scientific inquiry.

Freedom of Scientific Inquiry

It is not an ancient freedom. Commencing about 400 years ago, this concept developed slowly with the growth of science itself. Hopefully, the day is past when anyone would seriously argue that the Government should prohibit free expression of new ideas simply out of fear of the ideas themselves. Totalitarian governments may fear new ideas, even as they fear their own people. But in the United States, the freedom openly to espouse and debate new ideas has been the essence of liberty and the guardian of democracy. That great principle of American political history is embodied in the first amendment. But where did the first amendment come from?

Man's concept of freedom has been profoundly influenced by the search for truth, the single commitment of science. The case of Galileo, the consequences of Darwin are milestones in the history of our civilization. Those freedoms enshrined in the first amendment--freedom of speech, of religion, of the press, of assembly--came to be cherished precisely as the power of science freed mankind from dogmatic religious and political thought. Scientific inquiry has challenged the dogma of an authoritarian world for the last 400 years; it has freed men's minds as it eased their toil. After two centuries of science, it was Thomas Jefferson who said, "There is no truth on earth that I fear to be known." And I hold that it must certainly be more dangerous for a people to live in ignorance than to live with knowledge.

As Jacob Bronowski noted,

Science is the creation of concepts and their
exploration in the facts . . . First comes
independence in observation and then in thought
. . . Dissent is the native activity of the
scientist even though it has gotten scientists
into much trouble over the centuries. And
it is this value which science has given to
all of our society, for dissent is also native
in any society which is still dynamic and
growing. It is not an end in itslef. It is
the mark of a deeper value. It is the mark
of freedom . . .*

Specifically, it is that freedom that is protected by
the first amendment.

Freedom of inquiry, therefore, while not explicit, is
surely implicit in the first amendment. As the U.S.Supreme
Court said in Griswold v. Connecticut:

The right of freedom of speech and press
includes not only the right to receive, the
right to read . . . and freedom of inquiry,
freedom of thought, and freedom to teach . . .

For scientists, experimental research is an inseparable
aspect of free scientific inquiry. Such research, however,
is not merely words or the communication of thoughts, it
involves action. Research is empirical observation and the
testing of hypotheses against the hard reality of experience.
And while the protections of the first amendment can extend
to actions as well as words, the extent to which research, as
action, is thus protected under the first amendment is an
issue which is untested in the courts.

To be sure, first amendment coverage cannot confer an
absolute right of freedom from all restraints. We have
readily accepted various such with respect to scientific
inquiry, as, for example, those regarding the use of human
subjects. But we need accept no constraint other than those
found absolutely essential to protect against injury other
values that we cherish.

Importantly, what the Government may not prohibit, it
may nonetheless choose not to support financially. For the
greater part of basic research in the United States today,
that would be akin to prohibition.

---

*J. Bronowski, Science and Human Values (NY: Harper and Row,
1959): 77-79.

Thus the potential power of the Government to limit or inhibit the development of new knowledge is vast.

We must, therefore, be ever vigilant against the argument that the Government must decide what is safe for the American people to learn and, hence, to know, or that the Government should be the arbiter of knowledge. In this sense, it is not so much the courts as the Congress, by its authorizations and conferral of regulatory authority, that is keeper of the first amendment.

The objective of some who have proposed the regulation of recombinant DNA research is to use the power of Government for the suppression of ideas that may otherwise flow from such research. That would take us back to an era of dogmatism from which mankind has only recently escaped. And it would be a feckless task. In the long run, it is impossible to stand in the way of the exploration of truth. Someone will learn, somewhere, sometime.

As for the specific concern that research on recombinant DNA today will lead to genetic manipulation of the character of human beings at some future date, if it be true, and I am very skeptical, there is a long and extraordinarily difficult road ahead before that might become possible. Those who harbor this fear join that fraction of our fellow citizens who, as Gerald Holton has remarked, have come to "consider science as the serpent seducing us into eating the fruit of the tree of knowledge, thereby dooming us."

Although it is sadly true that the ingenuity of man seems never to be greater than when engaged in the skills of weaponry, that is scarcely the matter with which we deal. Happily, this technology which offers promise of great boon to mankind, affords little obvious prospect for its misuse in the hands of evil men.

Unfortunately, I know too well that the latter promise may prove to have little validity. The fruits of all research really are unpredictable; in the present instance one should certainly withhold judgment until such time as some vision of future evil may appear to have a more tangible basis in fact than anyone can present at this time.

It is understanding of the genetic process that is the principal objective of the research we discuss. But the practical application to be explored is not to tamper with man's genetics, it is to repair them, if and when that proves possible.

The technique of recombination of DNA has already proved its power to illuminate for us the nature of the genetic apparatus which is the essence of life. And there is much, much more to be done. Accordingly, I beseech the power of the first amendment by asking the Congress to abjure any legislation which, by accident or design, will limit the freedom of inquiry, save for express and compelling cause as elsewhere instructed by the Constitution.

This real issue before the Congress is to insure adequate protection for the public from any hazards that may result from recombinant DNA research. It is to that question that I now turn.

Protection For the Public

The circumstances are extraordinary and, I suggest, rather puzzling. We have witnessed an outpouring of concern that has brought us to the point of serious contemplation, by the Congress, of legislation intended to avert a risk which, today, remains hypothetical.

Since first the matter arose, I had had difficulty in imagining that this research could engender any risk greater than that which is daily accepted by the tens of thousands of individuals who minister to those afflicted with genuine viral or bacterial infection. Since the time of Koch, every day, the world over, thousands of technicians with a modicum of training in sterile technique, have cultured and otherwise manipulated genuine, virulent pathogenic bacteria taken from the urine, stool, sputum, and blood of patients.

And they have done so, all things considered, with a remarkable overall safety record. That risks exceeding those of everyday, unregulated, routine hospital practice may attend any experiment with recombinant DNA that a knowledgeable investigator might reasonably wish to perform has never been made clear to me. Nor are the viruses to be utilized in these studies manmade objects. They are abroad in nature; they have been harvested and studied, with neither P-3 nor P-4 facilities, largely without incident. I have difficulty in believing that a fragment of such a virus, or indeed, an entire viral genome, placed in an innocuous bacterium, transforms it into a raging beast unmanageable by standard, cautious laboratory procedure.

Thus, the history of this subject has laid upon those who would conduct this research the strange burden of offering a positive proof of safety, when the extraordinary risks they must show not to exist are entirely hypothetical in the first instance.

I need not repeat or analyze the categories of risk which have been contemplated. These are enunciated clearly in the NIH guidelines and in the explanatory statements that have accompanied them. At one extreme, there are indeed experiments which anyone would consider beyond the pale, such as deliberately introducing the gene for resistance to penicillin into a virulent strain of streptococcus. Even under rigorously controlled conditions, such an experiment is unthinkable.

But it is also pointless and, hence, I cannot see that it is cause for concern. Moreover, anyone who would do that, or introduce the gene for botulinus or cholera toxin into an ordinary enteric organism would probably not be deterred by any regulations to the contrary.

As one considers the classes of experiments permitted under the guidelines such as, for example, those for which P-2 and even P-3 precautions are presently deemed necessary, one can easily raise cogent arguments to the effect that, in practice, very little risk can actually attend them. Nevertheless, I quite agree that until more evidence is in hand, all research in this field should be conducted as specified by the NIH guidelines.

Having adopted that position, the meaningful question is: If recombinant DNA research is conducted within the NIH guidelines, is the public health adequately protected? My response is definitely affirmative. Let me explain.

In response to a request from Congressman Paul Rogers, chairman of the Subcommittee on Health Affairs of the House Committee on Interstate Commerce, that the Academy tender its views with respect to these matters, our Assembly of Life Sciences appointed a Panel on Risks and Benefits of Recombinant DNA Research. The Panel consisted of distinguished members of the Academy, knowledgeable in molecular biology, genetics, infectious disease and epidemiology, chaired by Maclyn McCarty of the Rockefeller University, an authority on infectious disease, who is a member of the trio who first demonstrated that DNA is the stuff of which genes are made.

I will not attempt complete answers to the list of questions addressed by that panel. I will just give you a few examples.

Could the use of E. coli, a known inhabitant of the human intestinal tract, as the host organism for recombinant DNA experimentation lead to epidemics?

The answer, very unlikely.

Could recombinant DNA harbored in E. coli be transferred to other organisms in the intestinal tract?

That has been tried and failed.

Could novel organisms be created that could be ecologi-
cally disruptive or cause human or animal disease?

There have been a few tries to demonstrate that. They
failed. The answer is, extremely unlikely.

Are physical and biological barriers used in the experimentation adequate to prevent the escape of hazardous
organisms from the laboratory?

More than adequate.

Could genes from eukaryotic cells be transferred into
prokaryotic cells and express their genetic information in
ways that could be harmful?

Yes; they can be transferred, but there is absolutely no
reason to think them then harmful.

And finally, is it possible to develop for use in
experimentation an organism so fastidious in its growth
requirements as to be unable to grow outside of the laboratory?

Yes, and this has been done by Dr. Curtiss but in fact,
we deem it unnecessary to improve on Escherichia K12 which
is already in wide use. It is itself a safe organism.

So the principal conclusion of the report is that:

>   Currently available evidence leads us to conclude
> that the many benefits of recombinant DNA research and
> technology can be achieved with negligible risk to the
> biosphere when the work is carried out within the NIH
> guidelines.

They state further that:

>   The guidelines are conservative in their recom-
> mendations for the growth and containment of
> organisms containing recombinant DNA; currently

functioning arrangements provide flexibility to
permit future reductions or increase in control,
as appropriate, when new data become available.

Increasing experience has diminished the level of concern among responsible scientists concerning the level of
risk that attends this form of research.

Despite the distractions of the continuing debate on
this subject, those engaged in this research have been making
extraordinarily rapid progress. Accordingly, some of the
evidence that influenced our Panel has been only recently
collected. Let me briefly summarize their report.

Of considerable importance have been demonstrations that
genetic material may be transferred between rather significantly different bacterial species. For example, there has
been demonstration of transfer of plasmids* obtained from
Staphylococcus aureus into the very different Bacillus
subtilis; the plasmids were those bearing genes for
tetracycline and chloramphenicol resistance. In the author's
words:

> The demonstration that replicating plasmids
> are shared among species of bacteria as widely
> diverse as a Staphylococcus and a Bacillus or, in
> a different experiment, between Escherichia and
> Agrobacterium makes it likely that plasmid
> sharing occurs commonly in nature. This is
> pertinent to our views of natural microbial evo-
> lution and to concern for the uniqueness of
> constructing DNA recombinants in the laboratory
> which is a premise of current policy discussion.

---

*Plasmids are miniature chromosomes, circlets of double-
stranded DNA, present in bacteria in addition to their main
single chromosome. Like the latter, a plasmid replicates
each time the cell divides. Expression of its genetic information confers additional properties to the cell but is not
essential to cellular life except under special circumstances.
Plasmids are a genuine source of concern. "R" plasmids
bearing genes for resistance to antibiotics confer a selective advantage to cells containing them when antibiotics
are present. The threat, however, is the spread in nature
of antibiotic-resistant pathogens in consequence of the use
of antibiotics for therapeutic purpose in man.

Indeed, there is persuasive evidence that such transfer, among unrelated bacterial species, does occur spontaneously in nature--by conjugative transfer of plasmids, by uptake of bits of free DNA, and by movement of what are now termed "transposons." Genes for antibiotic resistance appear to be passed, for example, among enteric organisms and such pathogens as Hemophilus influenzae and Neisseria gonorrheae. The common crown gall tumor of cotlyedonous plants results from transfer of a large plasmid from Agrobacterium tumefaciens into the plant's own chromosome. These observations led Elwell and Falkow to state that:

> . . . while committees of scientists and laymen banter about recombinant DNA around conference tables, Nature . . . has been conducting experiments prohibited under the NIH guidelines for recombinant DNA research.

These observations take on added meaning in the light of recent experiments by Stanley Cohen that demonstrated that when bits of mouse DNA were taken up by cells of Escherichia coli, they were trimmed to size and built into an already existent plasmid within the living host cell by exactly the same kind of enzymatic "cut and paste" procedure that scientists have been using to introduce a foreign DNA into plasmid DNA before reinserting the plasmid back into its host Escherichia coli. Thus, the laboratory technique, using enzymes to make "recombinant DNA," appears to be nature's own technique for accomplishing the same kind of chemical gene transfer. The full extent of such spontaneous transfer of genetic information among species in nature remains to be ascertained. That relatively distinct, definable bacterial species remain the norm; that is, that all bacterial species do not become homogenized into one, is presumed to reflect the fact that in the vast bulk of such instances of incorporation of foreign DNA into a bacterium, the consequence is disadvantageous; almost all manmade recombinants grow very poorly even in rich culture media. We shall certainly learn much more about such matters in the future.

Particularly comforting to the panel with respect to the safety of such research within the guidelines, is the demonstrated inability of the Kl2 strain of E. coli--much less, presumably, K2--to colonize the human bowel. Moreover, this strain of E. coli lacks cell wall lipoplysaccharides, lacks the virulence factors that permit attachment to the lining of the intestine, and lacks the genes that cause some wild type enteric organisms to secrete the toxins that exact "Montezuma's Revenge." Even were it or its modifications to escape, E. coli Kl2 cannot take up residence in the human intestine. Hence, the argument that danger arises from the fact that E. coli is a normal, enteric organism has no basis in fact when the usual laboratory strain, K12, is employed.

Each real pathogen possesses an exquisitely balanced set of genes that enables it to occupy a unique econiche. The panel concluded that it seems extremely unlikely, perhaps impossible, to convert E. coli K12 into a pathogen merely by insertion of the amount of genetic information permitted by the recombination procedure. The increment in DNA can be no greater than 0.1-0.2 percent of the host cell genome, only a few genes, and that seems simply insufficient to the task. Yet this was among the principal hypothetical risks that had been posited.

The panel was impressed, as well, by the fact that, both in the general national experience and that at Fort Detrick, less than 1 percent of all individuals who have been infected in the laboratory with a pathogenic organism have transmitted their infection to another individual. Almost all cases involved airborne organisms whereas E. coli is waterborne. Moreover, never has such infection spread, secondarily to more than one other person. Epidemic, therefore, seems extraordinarily unlikely.

Taken all together, the panel's conclusion seems entirely justifiable: research with recombinant DNA, performed under NIH guidelines, offers negligible hazard.

Potential Applications

Let me turn for a moment to the exciting progress along that research trail which may lead to practical applications. As you will be aware, the structural gene that specifies rat insulin has been successfully introduced into E. coli. The genetic controls necessary to switch on the use of this information were not introduced at the same time and so, although the gene is reproduced with each cell division, it finds no expression. These cells do not make insulin or its precursors. But it is a good start.

Even more exciting is the quite recent report, as yet unpublished, concerning incorporation into E. coli of the structural gene for the small protein called somatostatin together with all the necessary information to make this system operative.

This is a scientific triumph of the first order; and the practical promise held out by this field of research is closer to realization. Although somatostatin will undoubtedly prove to have diverse therapeutic applications, this is probably not the technique by which it will be made commercially.

It is a relatively small molecule, hence can be made by the techniques of synthetic organic chemistry in substantial quantity and reasonably cheaply. The significance of this accomplishment lies, rather, in its clear demonstration that all the genetic information required to make a protein that is normally made in a eukaryotic cell, indeed in the brain of man, can be transferred to a prokaryote, and there operate as in its former environment.

Once again I must offer a caveat. From this historic point to the practical production of insulin, or growth hormone or that of any of the enzymes that might be useful for repair of genetic disease, or to production of antibodies or clotting factors, is still a very long distance.

These are much more complicated proteins for which simple spinning out of the strand of amino acids is insufficient. Each requires additional processing by various kinds of enzymic procedures which may not be normal to E. coli.

And it may be an even farther distance to the useful incorporation, by this procedure, of nitrogenase into corn or wheat. A great deal remains to be done before there can be any true practical payoff. But the first giant steps have been taken.

Let me repeat. The principal objective of this research is not the direct, immediate applications that can presently be visualized, however those may turn out. The objective is understanding of the structure and function of the genetic apparatus, for which recombination is a technique of extraordinary power. The ultimate rewards of such knowledge are beyond the horizons of our imagination.

Role of Government

Let me turn finally to the question of how the Government should position itself in these regards. In so doing, I speak only for myself; our panel and our Committee on Laboratory Biohazards have not taken a position concerning these matters.

As I stated earlier, with most of my colleagues I consider it wise to have the NIH guidelines remain in force, noting again that the current guidelines are more than adequate to protect the persons who are engaged in this work and to protect the public health generally. I consider it important that these guidelines be revised regularly, revised or strengthened in the light of accumulated experience.

While in force, the NIH guidelines should be applicable to all research, wherever conducted and regardless of its auspices, public or private.

I fervently hope that, to this end, there will be established no more than the bare minimal arrangements that would suffice to maximize compliance.

Although many responsible and knowledgeable scientists find no need for either legislation or guidelines, including some whom I judge to be best informed, others with whom I have discussed the matter have come to the reluctant conclusion that some legislation is required to assure that the guidelines are universally applied.

If the Secretary of HEW already has sufficient statutory authority for this purpose, so that no additional legislation is necessary, I would be particularly pleased to have him operate under such and eschew any additional legislation.

In any case, what additional stipulations are important? If some form of licensure is deemed necessary to assure the public that there is general compliance to those guidelines, and I am rather dubious to that requirement--I urge that licensure be limited to the operation of P-4 and P-3 facilities.

It is abhorrent and, in any case, impossible for the Federal Government to police all establishments conducting this kind of research and I would hate to see NIH burdened with such responsibility.

Responsibility for compliance must, as has frequently been suggested, rest with the responsible research-performing institutions, operating through their Committees on Biohazards.

For my part, I have confidence that industrial firms, with their acute sensitivity to exposure to liability claims, will voluntarily comply with the guidelines. I suggest that, like universities, each industrial organization should have a Committee on Biohazards on which some of the members should be drawn from the external community; there is a problem.

If the "public members" of their boards of directors

were so to serve, that might avoid the problems raised by proprietary information. That's not an adequate solution, but it is the best one I have thought of in that regard.

In large measure, the monitoring of this system can be accomplished without policing inspectors. Scientific journals can be expected to decline papers when the work described violates the guidelines; such journals now insist on an appropriate statement concerning the conditions of human experimentation.

Work performed under research grants and contracts is done under a voluntary memorandum of understanding and agreement with the Government: violation can result in loss of supporting funds.

Commercial firms can be denied patents or marketing licenses, or be subject to cease and desist orders under the authority the Secretary probably has.

For operators of P-3 and P-4 facilities the most meaningful penalty would be loss of license. These are all more effective than would be a system of formal penalties. In any case, I consider criminal penalties for noncompliance to the guidelines to be highly inappropriate.

Public discussion of this research, of its intellectual and potential practical applications, will undoubtedly continue; and the degree of risk will probably remain a subject of debate.

It has been an extraordinary drain on the energy and time of scientists so engaged--some in the room--to defend the conduct of such research in each of the States and cities where this question has arisen. To the extent that such risks have reality, it is unlikely that they are greater or more threatening in one locality than in another.

There is no rationality in rendering the conditions of such research of varying regulatory stringency among the thousands of separate jurisdictions in the United States. Accordingly, whatever the agreed upon Federal regulations, they should be applied uniformly and without exception across the country.

Should any legislation in this regard be enacted, it should most certainly include a sunset clause to assure its expiration in due course, unless there be compelling reasons otherwise.

The Panel on Risks and Benefits of Recombinant DNA Research believes that the body of evidence acquired or adduced over the last few years completely indicates that DNA research performed under the NIH guidelines presents no real risk to the public health.

Past experience teaches that no manifest epidemics have ever arisen from laboratory work even with extremely pathogenic or contagious organisms and recombinant DNA work uses nonpathogenic organisms.

Knowledge of the biology of the E. coli K2 hosts gives confidence when the research is in the hands of trade workers, no untoward events will concur.

We conclude with the majority of scientists that many of the concerns expressed in 1973 have been satisfactorily answered. Several scientists who first called for extreme caution have stated their belief that research under the current guidelines is safe.

Indeed, it is probable that most recombinant DNA work using enfeebled E. coli systems could be carried out with safety without anyspecial precautions whatever.

Risks of Over-regulation

Consequently the public and its representatives in legislative and administrative offices should consider carefully the risks of overregulating this kind of endeavor. Recombinant DNA research under the guidelines has such great promise for rapid future benefits and so little chance of causing harm that its regulation should be implemented with a minimum of restriction.

Cumbersome and punitive legislation is not needed. The financial cost of overly cautious containment and enforcement, the delay in achieving benefits and the penalties incurred by restricting freedom of inquiry are real risks to be considered in setting up regulations.

Simple administrative procedures, uniformly applicable throughout the Nation with a minimum of redtape should be devised to permit work to proceed under the guidelines. The guidelines themselves should be under continuous scrutiny and revised when necessary, more expeditiously than in the past, so that the recommended procedures can more realistically reflect the increasing evidence that there are no practical risks from recombinant DNA research under the guidelines.

The principal argument in favor of the legislation is to
extend the force of the NIH guidelines to research conducted
under other than Federal auspices. And it is important that
that be achieved.

Yet I find it something of an embarrassment that the
Congress should be entertaining legislation designed to protect the American people against alleged extraordinary risks,
the existence, nature, and magnitude of which remain a
matter of speculation.

In view of the considerations and experience that I
have summarized and the absence of any untoward incident to
date, it may seem not unreasonable to delay legislation,
giving opportunity for yet further relevant observation by
workers in this field while placing upon NIH the obligation
to design and perform, under suitably controlled conditions,
a series of "worst case" experiments that may afford greater
appreciation of the true magnitude of the risks that have
been postulated.

The risk of violating that which we cherish under the
First Amendment surely loomsas large as the hypothetical
risks against which these safeguards are being erected.

And I am concerned that each instance of regulation of
research will facilitate the next.

An excess of zeal to protect us against all risks, however minor, particularly when seen against the backdrop of
our chancy world, could seriously cripple science, the principal tool our civilization has fashioned to mitigate the
condition of man.

The research with which we are here concerned was begun
in our country. It has rapidly been adopted wherever
biological science is pursued and the world is watching how
we shall now conduct ourselves.

Early in this history the very individuals most successfully so engaged were those who "blew the whistle." In a
unique event, they restrained their own endeavors. No
discussion has ever been more open and uninhibited than what
followed and the Government has certainly responded with
deliberate speed.

Had we to do it over again, I would have made sure that
Paul Berg's committee was appropriately weighted with some
clinical scientists with long experience in epidemiology and
in the handling of genuine pathogens. We might then have

moderated the excessively strident dialog that ensued

But I rather suspect that we would, nevertheless, have been gathered here this morning and that the state-of-the-art would be much as it is today.

In his book on Congressional Government (1885) Woodrow Wilson said, "If there be one principle clearer than any other it is this: That in any business, whether of government or mere merchandising, somebody must be trusted."

I suggest that, with respect to the factual aspects of technical matters, the somebody must be the group of knowledgeable experts.

Sir Thomas More said in 1516 that "the singleminded man must not govern, but teach," Since, 20 years later, he went to the scaffold for neglecting his own counsel, I, for one, am pleased to leave to the Congress its decision.

10

The Recombinant DNA Controversy
A Model of Public Influence

Harold P. Green

It has been fashionable for science policy analysts to discuss the recombinant DNA issue as involving an unprecedented effort to impose government regulation of scientific research. This is simply not true. Similar issues have arisen in the past, notably in the case of atomic energy in 1946. For example, consider the position of Dr. John von Neumann when he testified before the Special Senate Committee on Atomic Energy on January 31, 1946:

> It is for the first time that science has produced results which require an immediate intervention of the government. Of course, science has produced results before which were of great importance to society, directly or indirectly. And there have been before scientific processes which required some minor policing measures by the government. But it is for the first time that a vast area of research, right in the central part of the physical sciences, impinges on a broad front on the vital zone of society, and clearly requires rapid and general regulation. It is now that physical science has become 'important' in that painful and dangerous sense which causes the state to intervene.
>
> Considering the vastness of the ultimate objectives of science, it has been clear for a long time to thoughtful persons that this moment must come sooner or later. We now know that it has come.

The legislation on atomic energy represents

Reprinted by permission of the <u>Bulletin of the Atomic Scientists</u>. Copyright © 1978 by the Educational Foundation for Nuclear Science.

the first attempt in history to regulate science
in this sense. In past wartime and peacetime
emergencies, governments did influence various
phases of the social effort, including science,
in order to promote military or economic ends.
However, such efforts were always limited in
time and in scope, and directed toward some
ulterior, independent purpose. It is only
now that science as such and for its own sake
has to be regulated, that science has out-
grown the age of independence from society.

Many scientists regret this, and I am one
of them. Atomic physics in particular is now
losing a good deal of its detachment and abstract-
ness, and will probably never again be the same
as before 1939. I repeat: From the scientists' special viewpoint, this evolution is
probably not a desirable one--but nobody can
change it, and we must admit that it is taking
place. There is clearly a need for the
Government's intervention here and now.[1]

Congress went on to enact the Atomic Energy Act of 1946,
which imposed stringent regulatory controls--still in
effect--on scientific research in the atomic energy area.

Of course, there are important distinctions between
the base of atomic energy and that of recombinant DNA.
There are also important analogies. In both, scientific
experiments and resultant technology have the potential
for both enormous benefit and catastrophic disaster. In
both, painstaking care and stringent regulation can drastically reduce the probability that a catastrophic disaster
will occur. Both, therefore, have the capability of a
low-probability, high-consequence accident. In both areas,
scientists and technologists create products that do not
exist in nature, and protection of the public health and
safety requires techniques to contain these products in
physical structures so as to prevent their entry into the
general environment. In both areas the health, safety,
and security of the public rest ultimately upon faith in
the omniscience and infallibility of the human beings who
design and implement the scientific and technological
endeavors and the safeguard systems within which these
endeavors are conducted. Finally, both raise substantial
ethical and moral issues.

It is also worth noting that, just as the scientists
engaged in recombinant DNA took the initiative in 1973 and

1974 in alerting society to the social implications of their
work, so the atomic scientists in 1945 organized a major
effort to bring about effective social controls over atomic
energy.

There are two ultimate questions involved in discussing
public policy decision-making for recombinant DNA science
and technology: (1) How should government assess benefits
and risks in deciding the extent to which recombinant DNA
activities will be funded by the government? (2) To what
extent should DNA activities, including those that are
privately funded, be regulated by government? These are two
separate questions, and regrettably space does not permit
my discussing some important observations that should be
drawn in analyzing these two decisional modes. Suffice it
to say at this point that government seems inclined to
weight risks more heavily when it views activities that
are privately funded than when it considers restraints on
funding its own programs.

Nor is this the occasion for discussion of the intri-
cacies of risk-benefit assessment. It is clear, however,
as one looks at recombinant DNA activities--as in the case
of all young scientific and technological programs--that
the promised benefits are obvious and relatively immediate
while hazards tend to be speculative and relatively remote.
Moreover, since risk is the product of the magnitude of
potential adverse consequences multiplied by the proba-
ability of the event, it is always possible to produce a
very low product when one of the factors is a number
approaching zero. Thus, when it is assumed that human
caution, coupled with stringent regulation, makes the likelihood of a serious mishap vanishingly small, this auto-
matically makes the risk very small, even where the
potential disaster in the event of such an occurrence is
extremely large.[2]

Scientific objectivity, coupled with the scientists'
propensity for optimism that human beings can reliably
ensure accomplishment of what is possible, generally tends
to produce assessments that risk is very low. This certainly has been the case in nuclear power where we are
told there is no need for the public to be told about the
magnitude of potentially astronomic catastrophe that
inheres in nuclear power plants because of the asserted
vanishingly low probability that multiple events will
occur to trigger a major accident. Parenthetically, I
might note, this optimistic assessment is performed in
the face of the fact that the nuclear power establishment
has insisted upon enactment and re-enactment of the

Price-Anderson Act which limits the aggregate potential
liability of everyone who may be liable on account of a
nuclear power accident to $560 million, a figure that in
itself would represent the most massive disaster ever
caused by a peaceful man-made technology.[3]

This mode of risk-benefit assessment, as a predicate
of public policy decisions, requires the interested public
to accept the omniscience and infallibility of scientists
and technologists in constructing safe and fail-safe
systems. It is, moreover, an unnatural way to look at
risk-benefit assessment from the perspective of the public.
Most of us, I believe, when we ponder the question of
risk as laymen, give more weight to the magnitude of
adverse consequences than to probability of the occurrence.

Nevertheless, our society cannot permit itself to be
paralyzed by hypothetical possibilities of even astro-
nomically large catastrophes that may lurk in the shadows
of an uncertain, unknown, and unknowable future. It is
impossible ever to prove a negative--for example that recombinant DNA experiments will not result in a plague. Public
policy decision-makers must necessarily dec ide whether
particular levels of risk will be assumed in order that
society may enjoy desired benefits. The beginning of
wisdom is the recognition that no one--no matter how wise
and objective--has the capacity to state reliably and in a
universally acceptable manner what benefits the public truly
wants and what costs (and risks) it is willing to bear for
the sake of having those benefits. The best that we can
do is to ventilate the benefits and risks openly and can-
didly so that public sentiment can somehow, albeit
imprecisely, be reflected in the electoral, legislative,
and political processes to instruct, or at least signal,
public officials. In this respect, I believe the recombinant DNA example represents a model of responsible
public policy decision-making for science and technology.

The most important fact about the recombinant DNA controversy is that the group, (who I shall hereafter refer
to as the "organizing group") of scientists who called for
the moratorium and who steered Asilomar, chose to "go
public" with the problem. Whether or not they consciously
intended and desired to engage the public and the non-
scientific disc iplines in consideration of the issues, the
drama of the moratorium and of Asilomar ensured that the
issues would ultimately become publicly controversial. I
say "ultimately" because the issue did not--after the
initial flurry of past-Asilomar press accounts--attract
very much critical lay attention. It was not until

scientists such as Geroge Wald and Robert Sinsheimer spoke
out, and until the environmental activists entered the fray,
that the issue became a matter of real public concern. Still
it was inevitable that it would.

There are some who deplore the fact that the recombinant DNA discussion has become publicly and politically controversial. To some scientists this is "washing dirty
linen in public." The fact is, however, that the public
and laymen generally become interested in technical matters
only when they become controversial. Controversy sparks
interest, concern, and study. It not only focuses wide
attention on the issues, but also focuses the major issues
so that they can be given direct attention. This is not to
say that all forms of controversy are per se desirable.
For example, the controversy over nuclear power has produced a paralyzing polarization in which no one can take
satisfaction. There are ways to keep healthy controversy
from becoming paralyzing polarization; and the key to this
is candor and openness on the part of the concerned scientists with a willingness to discuss and debate the issues
on their merits. Absent such candor and willingness the
controversy will degenerate, as in the case of nuclear
power, to a struggle between half-truths propounded on
both sides.

As suggested above, a major area of potential polarization concerns the societal acceptability of an activity
that has the potential for extremely catastrophic consequences. Defenders of such activities usually smother
discussion of possible catastrophe under a blanket of confidence that such an event, although remotely possible,
simply will not occur. They take comfort in the fact that
experience to date--even if only three or four years--with
the activity has not brought any real danger to light.
They attempt to demolish those who are fearful about potential catastrophe by arguing that there is no scientific
evidence that such a catastrophe is possible, let alone
likely. This kind of approach is evident in the recombinant DNA experience.

At Asilomar, for example, there was no discussion of
the potential adverse consequences that led to the moratorium in the first place. Instead all emphasis was placed
on the fact that the scientists' prudence, coupled with
judicious self-regulation could probably prevent an occurrence that might trigger such consequences. Now we are
told that in three years since Asilomar, new scientific
studies have indicated that the risk was "overestimate"
and that what was then perceived as "potentially

catastrophic and uncontrollable" is now regarded as only "potentially dangerous but controllable."[4]

Assessments of this kind tend, for obvious reasons, to have a counterproductive impact on the intelligent lay public which knows, at least intuitively, of Murphy's Law--that what can go wrong will go wrong. In addition, coupled, as they usually are, with the objective of protecting the public from undue alarm that might operate to frighten it into forgoing the benefits the scientists know it wants, these assessments tend to antagonize thoughtful laymen because of their "trust big brother because he knows best" overtone. It is easy to translate these lay perceptions into the ad hominem contention that the scientists are really trying to protect their own vested interests; and these contentions are exacerbated when it appears that government funding agencies are concerned with promotion and sponsorship as well as with protection of society against hazards.

There are some who suggest that the organizing group and NIH raised and dealt with the moral and policy issues of recombinant DNA science and technology for the primary purpose of being able to apply the proverbial whitewash. Having bravely and responsibly raised these issues in public, they and NIH proceeded--so some cynics say--to attempt to impose their own solutions that would pay lip-service to ethics and morality while permitting the scientists to continue their experiments with a clear conscience.

In my own interpretation of events, the organizing group raised the issue squarely and was quite prepared to accept whatever judgment the political process might render. It had hoped, although it actively sought the participation of non-scientists, to limit the scope of the debate to the area deemed relevant and to keep it within the bounds of what they regarded as scientific responsibility.

The organizing committee and NIH lost control of the process when scientists and environmentalists with more radical views entered the debate. As typically happens in political controversy, the discussion became shrill and strident, and some of the new partisans--even those who were eminent scientists--began to discuss the issues in ways that the scientific establishment regards as unscientific. It is said, for example, that some scientists were deplorably offering contentions for which there was no scientific evidence. As the organizing group and NIH came under attack for their alleged generosity in permitting and sponsoring recombinant research, they increasingly "hunkered down" into a defensive posture. Much too frequently, like the Atomic Energy Commission of the past and the Nuclear Regulatory Commission of the present, NIH's public pronouncements appear to potential critics to reflect a greater concern with advancing the science and technology than with protecting the public health and safety.

The real issue centers upon the time and circumstances under which a useful societal activity should become subject to regulation. Where is the burden of proof? Must those who favor regulation prove that the activity is harmful? We know that, more often than not, regulation is imposed, too little and too late, and only after great harm results from the activity. Increasingly, therefore, our political system has been moving toward earlier regulation that offers a hope of preventing injury in the first place. In the food additive and drug areas, for example, we do not permit substances to be used without some kind of demonstration that harm will not result. A major part of the present recombinant DNA controversy is based on the reluctance of scientists to see their ability to perform good deeds restricted on the basis of mere speculation, as opposed to scientific evidence of potential harm.

I am reminded of some wise words spoken in October 1965 by Admiral Hyman Rickover when he delivered an address on "A Humanistic Technology" before the British Association for the Advancement of Science in London.

> Though a technology may clearly harm paramount human values, restraining laws will not be forthcoming unless public demand is sufficiently vocal and persistent to wear down the opposition of those with a vested interest in the harmful technology. Opposition tactics follow a pattern that is monotonously repeated whenever the attempt is made to regulate a technology in the public interest. It is well to familiarize oneself with the pattern.

> I have mentioned efforts to confuse the issue by arguing as if a law of science were at issue when in fact the proposed legislation deals with technology, not science. If this argument fails, the need for the proposed law is then categorically denied. Warnings of scientists are rejected as 'unproven' and 'exaggerated." Later, when these prove to have been entirely correct, the argument shifts from the substantive question

of whether a technology is harmful to an attack on
the legitimacy of any kind of protective legislation. Such legislation would violate basic
liberties,it is claimed; it would establish government tyranny and subvert free democratic
institutions. If all this proves futile and
legislation is imminent, there will be urgent
demands it be postponed until 'more research' can
be undertaken to establish the appositeness of
the proposed law.

These delaying tactics are highly effective.
It takes firm commitment to a humanistic technology to push through needed tactics of opponents
as well as thorough understanding of the fili-
bustering tactics of opponents, and great skill
in combating these tactics. No wonder public opinion
and the law have nowhere fully caught up with
those who misuse technology. Often as not they
escape with impunity, no matter how gravely they
injure man or society.

Rickover's remarks were made in connection with the
thalidomide episode, and he suggested that it was the technologists (the pharmaceutical companies) who wore the
"black hats" and the research scientists who wore the "white
hats." Although in the recombinant DNA controversy, it is
primarily scientists on both sides, Rickover's remarks
describe the situation remarkably well. It is not science
as the pursuit of knowledge per se that it is sought to
regulate, but only applications of science that some fear
may injure man or society. We are told by proponents of
freer research that restrictions on recombinent experiments
will half the march of science. Warnings about the risks of
recombinant DNA experiments in technology are rejected as
totally lacking in supporting scientific evidence even when
they come from reputable scientists. Restrictions on
research, it is argued, violate the Constitutional guarantee
of freedom of scientific inquiry. And, finally, we are
urged, let's do more research or let's create a study commission to decide whether we really need regulation and, if
so, what kind. Arguments of this kind have, within the past
year, apparently chilled the prospects for Congressional
legislation dealing with the recombinant DNA problem.

The atomic scientists of 1945 and 1946 were willing,
perhaps eager, to bring their laboratories under positive
government regulation. On the other hand, there is fierce

resistance on the part of many scientists to any limitation on their recombinant DNA research. Why the difference? Probably because in 1945 and 1946, both the immediate bene- fits and the potential adverse consequences of atomic energy were seen as essentially military in nature. On the other hand, the biomedical research in which recombinant DNA techniques are used appear to be intrinsically and impor- tantly benign and beneficial. After all, one can hardly complain about research that offers hope of winning the war against cancer.

Curiously, however, although the controversy is osten- sibly over the question of regulation of recombinant DNA research, in reality it is largely over the question whether such research will be funded by the government and, if so, in what kinds of laboratories. Surely, government that funds scientific research because it is perceived as promising public benefit need not carry a heavy burden of justification to decline to fund research that is perceived as involving public detriment. And if it has the power not to fund the research at all, it clearly has the power to fund the research subject to compliance by grantees with specified conditions.

But even if the issue were one of regulation, our society cannot permit the scientists who practice recom- binant DNA science to decide whether and how they will be regulated anymore than it can permit pharmaceutical manu- facturers to decide whether and how their new drugs will be regulated. Science is not a sacred preserve immune from ordinary political processes.

Let me return to the case of atomic energy. Although the atomic scientists went public with the social issues in 1945 and 1946, the involvement of the public was not long- lived. Very quickly effective public discussion was smothered by a combination of security-imposed secrecy and a perception of impenetrable technical complexity, both of which were skillfully nurtured by the atomic energy establishment in order to permit decisions to be made by a small in-group rather than through the usual political processes. By the 1950's for example, respected senior members of Congress were saying such things with respect to pending atomic energy legislation as:

> It is very difficult for the members of the Senate . . . to pass upon a subject like the one being considered. . . ."

I have been unable to learn the merits of it,
and I think you will be unable to learn the merits,
because none of us are scientists.

Congress has no business legislating in this
field because it lacks the information necessary
to direct the Executive Branch intelligently.

Anyone who is not a scientist has to take the
word of the AEC and its scientists on requests
for money.

The present morass in which nuclear power technology finds
itself is a direct heritage of this calculated paternalism
which was finally shattered in the late 1960's with emergence of the environmental movement.

There is a real danger that recombinant DNA policy-
making will go the way of atomic energy. Obsessed with
concern that restrictions on research may in the short term
be more severe than is warranted by the hard scientific
facts, the recombinant DNA establishment is trying to maintain effective control over the decision-making process.

Such efforts, although in my view misguided and ultimately self-defeating, are in no sense evil. Indeed, the
effort is probably more subconscious, rooted in scientific
principle, than deliberate. Nevertheless, evidence of this
effort is seen in the insistence that public policy be
geared to what is today's scientific fact, avoiding any
speculation as to what scientific fact may be tomorrow; in
the rejection of arguments that seem to be based more on
emotion or feeling than on science; in lobbying efforts to
fight off more severe regulatory measures; and in the reluctance of the establishment to admit as serious participants
in the debate those who have "not done their homework" and
who, therefore, are incapable of speaking in the esoteric
vernacular of science.

Let me offer my own description of the arguments
through which scientists fight off regulatory actions.

First, there is no evidence that any harm has
resulted in the past or will result in the future.

Second, if some harm is apparent or likely, the
harm is trivial relative to the benefits.

Third, if harm is regarded as significant, we can
find a technological fix to eliminate, or to reduce the

level of, the harm.

Fourth, if nevertheless the harm appears to be
unacceptable, trust us, because as responsible scientists we simply will not do anything shown to be
injurious to society.

Fifth, if some scientists or technologists
persist irresponsibly in doing what is injurious to
society, that is the time for regulation, and we
will vigorously support it.

Such an approach is politically naive, and is hardly
effective in overcoming the typical syndrome of regulation
imposed only after too many people are hurt too long and
too badly. Given today's extremely rapid pace of scientific
advance and the escalating destructive (in both a physical
and ethical sense) potential of technology, it is imperative
that the need for societal controls be fully debated in the
political process in the very early stages of scientific
and technological development. It is usually too late to
impose effective control after unacceptable injury has been
demonstrated, because by that time too many people have
acquired a vested interest in the profit and prestige
flowing from the development, and also in the enjoyment
of its benefits.

Regrettably, there is no easy answer. Recombinant DNA
science and technology, like nuclear power, will gain long-
term acceptance only if the public feels confident that it
has been given all of the facts and permitted to make its
own judgments. Public perception that it must rely on the
expert judgment of scientists because it cannot comprehend
the issues may for a short time give carte blanche to
science, but ultimately it will breed a counter-reaction.
Science must find ways to communicate the significance of
its works to the public and to politicians in the language
of the ordinary political discourse. Science must be
removed from its pedestal and treated as any other societal
activity. Scientific activity, like any other activity,
should become subject to limitation whenever the public,
reacting through the political process, concludes that
scientific developments will injure or threaten important
interests or values.

Of course, opening up science to the vagaries of the
political process also opens the door to its potential
brutalization by know-nothings. But this is equally true
of every other societal activity. Science has in fact
occupied an exalted, almost untouchable, position in our

society for the past 30 years. It cannot, however, remain on this pinnacle in the face of the dramatic changes that have taken place in our society during the 1970's.

In 1919 Justice Oliver Wendell Holmes wrote in his famous dissenting opinion in Abrams v. United States (250 U.S. 616):

> The best test of truth is the power of the thought to get itself accepted in the competition of the market.

In the arena of public policy decision-making, scientific truth will not prevail if it is handed down from on high; it will prevail only if it finds acceptance in the competition of the market.

References

1. Hearing on S. 1717 before the Senate Special Committee on Atomic Energy, 79th Cong. 2d Sess. 206 (1946).
2. See Green, Cost-Risk-Benefit Assessment and the Law: Introduction and Perspective, 45 George Washington Law Review 901 (1977); Green, The Risk-Benefit Calculus in Safety Determinations, 43 George Washington Law Review 791 (1975).
3. Green, Nuclear Power: Risk, Liability, and Indemnity, 71 Michigan Law Review 479 (1973).
4. See testimony of Clifford Grobstein on September 7, 1977, Hearings before Subcommittee on Science, Research and Technology, House Committee on Science and Technology, 95th Congress, 1st Sess. 1043-1044 (1977).
5. H.P. Green and A. Rosenthal, Government of the Atom: The Integration of Powers, 78 n. (1963).

Part III
Regulation of Scientific Inquiry and the First Amendment

11

The Constitution and Regulation of Research

Thomas I. Emerson

The constitutional problems involved in governmental regulation of scientific research have never been directly addressed by the Supreme Court. Although numerous laws and regulations affect various aspects of such research, the far-reaching issues that are raised by current proposals for controlling the use of recombinant DNA technology present novel constitutional questions. My ideas on the subject are wholly tentative, and I reserve the right to change my mind. Moreover, my conclusions can be set forth here only in the briefest manner.

The primary constitutional provision applicable is, of course, the First Amendment. That fundamental guarantee has the broadest reach and imposes the strictest limits on the kind of governmental action we are considering here. Other constitutional requirements--including due process, equal protection, and perhaps the right of privacy--may also be involved. In general, however, these provisions of the Constitution perform a supplemental function here. They have an impact only where the First Amendment cannot be invoked. Hence they are largely limited in scope to the detailed issues that will arise after the basic framework of control has been shaped by the demands of the First Amendment. My discussion in this initial presentation, therefore, will deal exclusively with First Amendment issues.

I will also consider the questions in terms of control over recombinant DNA research.

There can be no doubt that the First Amendment provides extensive protection to freedom of scientific research. It declares that "Congress," and that term includes all branches of government, "shall make no law . . . abridging the freedom of speech, or of the press, or the right of the people peaceably to assemble, and to petition the Government

for the redress of grievances." Although phrased in somewhat
narrow and specific terms, the First Amendment undoubtedly
was intended to, and certainly has been interpreted to,
forbid the government to intrude upon all forms of expression.
It was designed to maintain an effective system of freedom
of expression in the United States. And freedom of scientific inquiry is surely one of the fundamental elements of a
system of free expression.

As to the intention of the framers of the Constitution,
I have recently had occasion to summarize their views with
respect to the function of the First Amendment in the
following way:

> The process is essentially the method of
> science. The theory of freedom of expression,
> indeed, developed in conjunction with, and as an
> integral part of, the growth of the scientific
> method. Locke, following Hobbes, based his philoso-
> phical and political theories on the premises of
> science. And the proponents of free expression were
> all men who, in the broad sense at least, put their
> faith in progress through free and rational inquiry.
> Hence the process they envisaged operates upon the
> same principles as those that guided the men of
> science: the refusal to accept existing authority;
> the constant search for new knowledge; the insistence
> upon exposing their facts and opinions to opposi-
> tion and criticism; the belief that rational discus-
> sion produces the better, though not necessarily the
> final, judgment. This process did not ignore prior
> knowledge or opinion, but it did insist upon the
> responsibility of the individual to challenge such
> opinion, and upon the obligation of all to make
> reasoned conclusions based upon the evidence.[1]

The Supreme Court has consistently applied the First
Amendment in accordance with this original intention. Over
50 years ago, before the First Amendment had been made
applicable to the States, the Court held unconstitutional a
State statute that made it a crime to teach languages other
than English in the public grammar schools, condemning such
restrictions upon the freedom of teachers to teach and of
students to learn as a violation of due process.[2] Subse-
quently the Court made clear that the First Amendment
embodied the basic principles of academic freedom. In
Sweezy v. New Hampshire, reversing a contempt citation for
refusing to answer questions before a legislative investi-
gating committee concerning the contents of a university
lecture, Chief Justice Warren declared:

The essentiality of freedom in the community
of American universities is almost self-evident.
No one should underestimate the vital role in a
democracy that is played by those who guide and
train our youth. To impose any strait jacket upon
the intellectual leaders in our colleges and
universities would imperil the future of our
Nation.[3]

This theme has been sounded again and again by the
Supreme Court. Thus in Keyishian v. Board of Regents the
Court, striking down a State loyalty program for teachers,
stated:

Our Nation is deeply committed to safeguarding
academic freedom, which is of transcendent value
to all of us and not merely to the teachers concerned. That freedom is therefore a special concern of the First Amendment, which does not
tolerate laws that cast a pall of orthodoxy over
the classroom.[4]

And in Epperson v. Arkansas, where the Court invalidated
a statute that prohibited teaching of the theory of evolution
in the public schools, it repeated:

Our courts . . . have not failed to apply
the First Amendment's mandate in our educational
system where essential to safeguard the fundamental values of freedom of speech and inquiry
and of belief.[5]

Thus we start with a strong commitment to the principles
of free inquiry and a heavy presumption against any form of
governmental interference.

First Amendment and DNA Recombinant Research

There are several ways to approach the more specific
problem of applying the First Amendment to governmental controls over recombinant DNA research. My own theory of the
First Amendment, which I call the full protection theory,
derives from the "absolute" position taken most prominently
be JusticesHugo Black and William O. Douglas. It holds
that one must first determine whether the conduct involved
is "expression," which is covered by the First Amendment,
or "action," which is not. If the conduct is found to be
"expression" then it is fully protected by the First Amendment against any form of governmental regulation or interference' if the conduct is "action" it is not protected by

the First Amendment against any form of governmental regulations or interference; if the conduct is "action" it is not
protected by the First Amendment, though any governmental
regulations must conform to the due process clause, the equal
protection clause and similar constitutional provisions. It
should be noted at once that the Supreme Court has never
accepted this full protection theory. Nevertheless I
believe it is the only sound analysis and that its use here
will throw a helpful light on the issues now before us.

The first question, therefore, is whether the conduct
involved in DNA research constitutes "expression" or
"action." It seems to me that the development or exposition
of theoretical ideas about DNA and other genetic materials
and processes is clearly expression. Such conduct involves
the search for truth in its primal form. The fact that the
researcher works physically with complicated equipment does
not deprive the conduct of its character as expression. In
similar fashion a telescope is used to study the stars, an
accelerator to study nuclear particles, a public address
system to carry on a public meeting, and a xerox machine to
make copies for distribution.

The more difficult question is the classification of
experimentation. Experimentation is a vital feature in the
development of new information, ideas, and theories. This
is particularly so in the physical sciences. One must conclude that it is often an integral part of scientific
research, that is, a part of the system of freedom of
expression. Analogous conduct is the marching in a demonstration, the publication of a newspaper, and the organization of a political party. Although all such conduct
involves more than sheer thinking or verbalization, nevertheless it is an essential feature of a system of free
expression.

On the other hand, at some point experimentation clearly
moves into the realm of action. Just as a political assas-
sination has an element of expression but is basically
action, so an experiment to test a theory of nuclear energy
which might blow up a city, or contaminate the atmosphere
of the whole world, is also predominantly action. The line
has to be drawn on the basis of all the facts in a particular case and in light of the proper function of a system
of freedom of expression in a democratic society.

On the basis of present information available to me it
is difficult to state more specifically what forms of
experimentation should be classified as expression, and what
as action. It does seem clear, however, that experiments

which pose a serious threat to the physical health or safety of a community, must be classified as action. Such conduct is analogous to the use of violence against persons or property in a demonstration, or the throwing of rocks through the windows of the White House. The physical element of the conduct is the paramount concern, and the conduct therefore falls into the realm of action rather than the expression of ideas.

On this analysis, the broad search for information about DNA, the formulation of hypotheses, the exposition and discussion of theories and methods would constitute expression, and be fully protected under the First Amendment. Thus the government could not prohibit, regulate or discourage in any way DNA research on the ground that mankind ought not to be pursuing ideas about ways to develop new forms of life. On the other hand experiments that presented a substantial and serious danger to the physical health and safety of the surrounding population could be subject to regulation without infringing the guarantees of the First Amendment. Only the requirements of due process, equal protection and other constitutional provisions would be applicable to such regulation.

Other Theories of the First Amendment

If we seek to ascertain the constitutionality of government regulation by more orthodox theories of the First Amendment, several possible doctrines are available. One is the classic clear and present danger test. Under this doctrine the issue would be whether the DNA research involved created a clear and present danger of a serious evil that the government had a right to prevent. For several reasons, however, the clear and present danger test does not seem to me acceptable. In the first place the Supreme Court has rarely employed the clear and present danger test in recent decades, and may be said to have abandoned it.[6] Secondly, as applied to the problems before us, the clear and present danger test would amount to little more than a general balancing of interests test. And, if balancing is to be employed, a more carefully structured balancing test, which will be discussed shortly, is available.

A second possible doctrine is the simple balancing of interests test. Under this doctrine the individual and social advantages of engaging in the DNA research contemplated would be weighed against disadvantages. The Supreme Court has applied such a balancing test in the past, and still continues to do so.[7] Nevertheless, as just observed, more sophisticated balancing tests have now come into use and would seem to be vastly preferable.

The orthodox doctrine most acceptable, and the one I
believe the Supreme Court would adopt, is a structured balancing test. According to this test, when fundamental First
Amendment rights are involved, governmental regulation is
valid only when the government sustains the burden of proving
(1) that there are "compelling reasons" for the regulation,
and (2) that the objective cannot be achieved by "less
drastic means," that is, by more narrowly drawn regulations
less detrimental to First Amendment rights. As the Supreme
Court said in Buckley v. Valeo, involving the constitutionality of the Federal Election Campaign Act:

> Even a 'significant interference with pro-
> tected rights of political association' may be
> sustained if the State demonstrates a sufficiently
> important interest and employs means closely drawn
> to avoid unnecessary abridgment of associational
> freedoms.[8]

The question then becomes, what constitutes "compelling
reasons" for governmental regulation of DNA research. Some
possible reasons can immediately be marked off as not compelling, in the constitutional sense, though they may be
compelling as the basis for decision by individual citizens.
Religious, moral or philosophical arguments that man should
not probe too far into the established order of nature would,
I think, fall within this category. For the government to
base controls on these grounds would run counter to the basic
premises of a system of freedom of expression. This, I take
it, is a lesson of such cases as Griswold v. Connecticut,
invalidating a State law prohibiting the use of birth control devices; Roe v. Wade, upholding the right to an abortion
in the early stages of pregnancy; and Stanley v. Georgia,
striking down a State statute which made it a crime to read
or see obscene materials in the privacy of one's home.[9] The
religious or moral views of one segment of society should not
be allowed to infringe upon freedom of inquiry.

From the other end of the spectrum, some reasons are
clearly "compelling." Experiments which can be plainly shown
to pose a serious physical hazard to the health or safety of
the community would be constitutionally subject to regulation.

In between lies a broad area which would be dependent
upon the facts demonstrated in the particular case. Thus
the degree of risk that would be needed to justify governmental regulations could only be determined in the light of

concrete information. And presumably the more serious the
risk the less degree of certainty that would be demanded. I
do not have sufficient information in my possession to move
beyond this degree of generalization.

The second portion of the structured balancing test
requires that the government regulation imposed be the
narrowest necessary to achieve the objective. This seems to
me to involve two kinds of limitations on governmental action.
One is that the governmental restrictions be kept to a bare
minimum. This would require, for example, that where
possible the control be temporary rather than permanent; that
where possible it be regulatory rather than prohibitory;
that it involve the least onerous burden; that licensing
or other forms of prior restraint be utilized only as the
last resort; and so on.

The other requirement of the least drastic means test,
in my opinion, is that the controls be imposed only from one
source, which must be the Federal government. The advantages
of decentralization in many situations are obvious. But
where delicate issues of academic freedom are involved, as
in the DNA research controversy, the fewer sources of governmental restriction the better. I think there is a little
doubt that a failure of the Federal government to preempt
this field would lead to serious and widespread infringe-
ments upon freedom of inquiry.

Importance of Adhering to
Procedural Standards

One further aspect of the First Amendment problem
remains to be noted. The Supreme Court, as a condition to
sanctioning legislation which impinges on First Amendment
rights, has usually insisted upon adherence to very strict
procedural standards. Thus it has held that restrictions can
be enforced against the exhibition of motion pictures alleged
to be obscene, against the holding of a meeting which may
result in violence, against the sellers of allegedly porno-
graphic books, and the like, only where procedures for
assuring adherence to First Amendment requirements are carefully maintained.[10]

In the case of DNA research, again, it is not possible
for me to spell out at this stage precisely what procedures
would be constitutionally required. In general they would
have to be the least burdensome compatible with workable
regulation. More specifically, two examples of the kind of
process necessary can be mentioned. First, some form of
rapid and effective court review, both of the regulations

issued and of individual decisions made under regulations, would clearly be mandated by the Constitution. Second, some procedure for utilizing experts and other non-partisan scholars in the decision making process, and for assuring that decisions will be made by institutions with a sensitivity for freedom of expression, would be essential. This is an area that should be given most careful consideration.

Conclusion

In conclusion, one can say that a democratic society is not incapacitated by the Constitution from protecting its vital interests so far as the development of scientific research is concerned. As a matter of fact, though the Supreme Court has not yet been directly involved, various forms of control are assumed or accepted as wholly legitimate by our society. No one would question that Nazi-type experiments upon human beings, no matter what their scientific value, are legally beyond the pale. The Atomic Energy Act regulates in closest detail the possession and use of certain substances, for scientific and other purposes, where unregulated activity might lead to public danger. Various drugs and other materials, useful in scientific research, are likewise controlled. Any actual physical dangers inherent in DNA research can be forestalled on the same basis.

Yet in doing this it is imperative that our long tradition of freedom of research and freedom of inquiry be preserved. For this purpose the First Amendment stands as a bulwark against small encroachment or massive attack. Regardless of what theory of the First Amendment is employed, the concrete results seem to be strikingly similar. The right to pursue knowledge and to expound ideas remains free. The right to engage in experimentation that physically imperils the health or safety of the community may be restrained. The difficult problem will be to maintain an appropriate balance between the two principles.

References

1. T. I. Emerson, Colonial Intentions and Current Realities of the First Amendment, 125 U. Pa. L. Rev. 737, 741 (1977).
2. Meyer v. Nebraska, 262 U.S. 390 (1923). See also Pierce v. Society of Sisters, 268 U.S. 510 (1925).
3. 354 U.S. 234, 250 (1957).
4. 385 U.S. 589, 609 (1967).
5. 393 U.S. 97, 104 (1968). See also Tinker v. Des Moines Independent Community School District, 393 U.S. 503

(1969); Healey v. James, 408 U.S. 169 (1972).

6. See Brandenburg v. Ohio, 395 U.S. 444 (1969). But cf. Nebraska Press Association v. Stuart, 427 U.S. 539 (1976).

7. See, e.g. Bigelow v. Virginia, 421 U.S. 809 (1975); Virginia State Board of Pharmacy v. Virginia Citizens Consumer Council, 425 U.S. 748 (1976).

8. 424 U.S. 1, 25 (1976). See also Sheldon v. Tucker, 364 U.S. 479 (1960); Hynes v. Mayor of Oradell, 425 U.S. 610 (1976); Shapiro v. Thompson, 394 U.S. 618 (1969).

9. 381 U.S. 479 (1965); 410 U.S. 113 (1973); 394 U.S. 557 (1969).

10. See, e.g. Freedman v. Maryland 380 U.S. 51 (1965); Carroll v. President and Commissioners of Princess Anne, 393 U.S. 175 (1968); Marcus v. Search Warrants, 367 U.S. 717 (1961). See also Speiser v. Randall, 357 U.S. 513 (1958).

12

The Boundaries of
Scientific Freedom

Harold P. Green

Two and a half years ago a group of scientists called
upon their colleagues throughout the world to establish a
moratorium on certain kinds of experiments involving recombination of DNA molecules. This moratorium, apparently
universally accepted, the subsequent NIH guidelines imposing
positive restrictions on such experiments, and prohibitions
suggested or adopted by various state and local governments,
have all stimulated discussion as to whether such restraints
in some way violate what has been characterized as "the
right to scientific inquiry." More specifically, it has been
suggested that scientists have a right to pursue knowledge
and this right is of the same dignity as freedom of speech
and of the press guaranteed in the Constitution of the
United States.

It is not surprising, therefore, that upon establishment of the AAAS Committee on Scientific Freedom and Responsibility, that committee would turn its attention in part to
the question whether there are in our American system of
government and law any boundaries to scientific freedom and,
if so, where these boundaries are to be found.

In my comments I intend to discuss these issues from my
dual perspective as a teacher of consitutional law and as a
student of public policy for science and technology. In
doing so, I shall not consider whether there are or should
be any limits on scientific freedom as a matter of morality
or policy, but only whether limitations are permissible as
a matter of constitutional law.

To begin with, I am not aware of any precedent or legal
authority that clearly supports the proposition that there is
a constitutionally protected right to pursue knowledge or to

Reprinted from Newsletter on Science, Technology, and Human
Values, June 1977, pp. 17-20, by permission of publisher.

engage in scientific inquiry. I believe, and I am prepared
to argue, however, that such rights are implicit in the
First Amendment freedoms of speech and press; and for purposes of this paper it is assumed that the Constitution
guarantees and protects such rights to precisely the same
extent as speech and press. Parenthetically, it seems clear
to me that a right to scientific inquiry can have no greater
constitutional dignity than freedom of speech. Let us
therefore explore the boundaries of freedom of speech in the
effort to understand the boundaries of scientific freedom.

It is impossible in the space allotted to me to give you
a complete exposition of the boundaries of freedom of speech
as enunciated in Supreme Court decisions. Suffice it to say,
some kinds of speech enjoy the protection of the First
Amendment; other kinds of speech do not. Even where speech
does enjoy such protection, the degree of protection is
variable. A distinction of crucial significance is that
between speech and action. Speech emanating from the vocal
chords is generally fully protected, but amplified speech is
not; one is constitutionally protected in cursing the flag or
a draft card, but he is not protected when he rips or tears
it; one is protected by the First Amendment when he engages
in vigorous debate with a foe, but not when he uses language
(fighting words) calculated to provoke a violent response;
one may discuss aircraft hijacking in his own home or office,
but not when he is sitting in a commercial aircraft.

Such precedents are helpful in drawing the constitutional boundaries of scientific freedom. Surely a scientist has
the freedom to think, to do calculations, to write, to speak,
and to publish. When, however, the scientist leaves the area
of such abstractions and turns to experimentation, he moves
within the range of action that may enjoy only some, or
perhaps very little or no, constitutional protection. To
the extent experimentation could be constitutionally protected, freedom would vary inversely with the degree of
perceived impact on persons and the environment. Thus,
where scientific research involves experimentation with
human or animal subjects or where it impinges upon the community, it would clearly become subject to regulation. It
is interesting to note at this point that Hans Jonas reaches
the same conclusion from the moral perspective. He tells us
eloquently, "The granting of freedom to thought and speech
...does not cover action, even if subsidiary to thought.
Action is always subject to legal and moral restraints."[1]

I think, so far as I have gone, scientists would sense
intuitively that what I have said is correct. They are,
after all, surely aware of a multitude of legal restraints
on what they can do and where, when, and how they can do it.

Where many would probably part company with me is on the question of where the burden of proof lies before government may properly restrict scientific freedom.

Again, the freedom of speech analogy is instructive. When we are operating in the realm of pure constitutionally protected speech--or abstract or theoretical scientific research--a proponent of restrictions must carry a heavy burden of proof. There must be a compelling governmental interest in the restriction (e.g., a clear and present danger to be protected against), and the restriction itself must be designed to intrude to the minimum extent possible on the constitutionally protected right. As, however, we move down the scale towards actions and experimentation, the burden shifts dramatically, and no more than a rational basis will be required to sustain the constitutionality of the restriction. For example, obscenity is not protected by the First Amendment. Therefore, it is not necessary for government to show a clear and present danger before it acts to restrict obscene speech. Obscenity may be prohibited without any showing that obscenity is harmful; indeed, it is not even necessary to show that the government actually thinks that obscenity is harmful; it is enough that government <u>may have believed</u> obscenity was harmful. This attitude reflects the currently prevailing judicial attitude that, at least where no constitutional limitation on goverment power is operative, the courts will not second-guess the legislature or executive as to the wisdom, desirability, or necessity for regulation. Thus, there has never been any doubt in my mind that a city's prohibition against recombinant DNA molecule experiments within city limits does not violate any constitutionally protected right of scientific inquiry where the city may rationally--even though perhaps not reasonably --have believed that such experiments might endanger the health and safety of the public.

At this point it is necessary to draw another kind of distinction--between government regulation of a scientific activity and a government decision not to fund that activity. We sometimes forget that government has no moral or constitutional duty to support scientific research, no matter how beneficial the hoped for results, and that no scientist has a constitutional right to have his research projects funded by the government.[2] I am reminded of Freeman Dyson's article in <u>Science</u> in 1965 in which he argued that NASA's decision not to continue funding Project Orion represented "the first time in modern history that a major expansion of human technology has been suppressed for political reasons." In the same vein, some scientists seem to believe that it is immoral or wrong if the government is really motivated by a fear that resulting scientific knowledge will be misused. Personally, I do not understand why if it is legitimate for government to fund research because it hopes for constructive knowledge it is illegitimate for government not to fund research because of concern that the resulting knowledge will be destructive. Indeed, as I have argued elsewhere, it is probably not realistically possible for our democratic society to impose timely and effective regulation over abuse of knowledge resulting from government-sponsored research and development.[3]

When government, for whatever reason, chooses not to fund a particular kind of scientific research, it is not interfering with scientific freedom. Scientists remain perfectly free to do this research if they can find the money elsewhere. On the other hand, a direct prohibition or restriction on scientific research may indeed represent an infringement of scientific freedom.

It requires only a moment's reflection to appreciate that there is really nothing new or novel from the standpoint of the NIH guidelines on recombinant DNA molecule experiments, or the Cambridge restrictions on such experiments. We all realize, or should realize, that government in the past has imposed restrictions on where and when certain kinds of research may be conducted. Obviously, city zoning laws may preclude experiments with explosives in the center of an urban population center, and it would probably be regarded as a legal nuisance if the explosives were experimentally detonated within earshot of the community at 2:00 a.m. We know that scientists are not free to experiment with human subjects or the fetus as they see fit. We know that there have been restrictions on the use of animals or cadavers in scientific research. We know that the Food and Drug Act and the Atomic Energy Act restrict and regulate certain kinds of research. We know that limits on the use of classified information may impede or bar certain scientific research programs. Indeed, Dr. Barry Casper, has raised the question of a moratorium on development of laser enrichment of uranium.[4]

It is not clear to me why, in the face of such precedents, the scientific community has become so edgy about scientific freedom in recent months.

When I began this talk, I made it clear that I would be discussing the boundaries of scientific freedom as a matter of constitutional law and not as a matter of public policy. It is important to distinguish clearly between these two concepts. For example, I personally would argue in favor of the constitutionality of the Cambridge prohibitions against recombinant DNA molecule experiments, but I would argue against such prohibitions on policy grounds. In recent years, we have become excessively accustomed to looking to the courts to protect what we perceive to be our rights, and we have lost sight of the fact that the first, and in many cases the only, line of defense of these rights is our legislatures. While an argument about a right to scientific freedom may be a useful piece of rhetoric in political debate, we should not take the existence of such right too seriously. In short, the principal point that I would leave with you today is that the boundaries of scientific freedom, at least in terms of current issues, are established primarily through the political process and are not rooted in consitutional law.

References

1. H. Jonas, "Freedom of Scientific Inquiry and the Public Interest," Hastings Center Report, 27 (1976): 17.

2. See, for example, my exchange with Bernard Davis, "Ethical and Scientific Issues Posed by Human Uses of Molecular Genetics," Annals of the New York Academy of Sciences, 265 (1976): 176.

3. See, for example, my paper "Law and Genetic Control: Public Policy Questions," Annals of the New York Academy of Sciences, 265 (1976): 170-175.

4. B. Casper, "Laser Enrichment: A New Path to Proliferation?" Bulletin of the Atomic Scientists, 33 (1977): 28-41.

13

Effects of Legal Regulation on Scientific Research Using Recombinant DNA Technology

David J. Newburger

My remarks are divided into three parts: First, I comment on some of the more specific topical matters that currently command wide attention. Second, I bring to your attention several aspects of the interrelation between regulation and innovation. Admittedly, these appear to be a random collection, but they represent concerns that are crucial for writing an effective regulation but often ignored in much of the important literature. The final part is devoted to an outline of the broadest features of a possible regulation of research activity using recombinant DNA technology. This part allows me to develop the applicability of the theoretical points outlined in the second part but does not purport to resolve the ultimate question of what form any regulation adopted by Congress should take.

I. The Topical Matters

In an effort to speak specifically to some issues without attempting to present an exhaustive analysis of the area, I have identified seven topics upon which I feel qualified to provide some comments.

1. What criteria should govern the decision to regulate recombinant DNA technology?

Very few public issues pose so knotty a problem as finding criteria for deciding whether to regulate research using recombinant DNA technology. On the one hand, we have absolutely no experience with injury to public health and safety resulting from this research and, therefore, no informed basis upon which to make a statistical evaluation of the risks involved. On the other hand, we can reasonably hypothesize that using this technology might result in the release of previously unknown organisms that can cause irreparable injury and against which there is no known

defense. The question comes down to whether we, as a
nation, are more risk averse to public health and safety
dangers than we are to dangers of impeding innovation through
unlimited research in the area. The most likely answer, it
seems to me, is that we are averse to both risks and will
attempt to steer a middle course. The only analogous cases
that comes to my mind are nuclear power, in which we are not
able to assess the potentiality of radioactive releases from
nuclear reactors, and fluorocarbons, in which we are unable
to assess the actual danger to the ozone layer of fluorocarbon releases. In both instances we have chosen to regu-
late--in the first, by requiring multiple safety systems
even when there may be wide agreement that fewer such
systems would suffice, and in the second, by restricting the
commercial use of fluorocarbons.

An important part of the search for criteria for determining whether to adopt regulation is finding those considerations should not be given weight. For example, it seems
clear that the decision should not be based on a nose count
of scientists who have expertise in recombinant DNA research
particularly or science in general, but have no claim to
expertise in discerning the public interest. It is true
that these scientists may be able to provide a basis for
defining the risks attendant with recombinant DNA research.
Such could substantially aid the direction of this subcommittee's deliberations. Care must be taken, however, to
distinguish opinions based upon rigorous research and those
that are personal to the speaker.

Further, the decision whether to regulate should not be
based upon the assertion that such regulation constitutes an
unwarranted governmental intrusion into academic freedom.
In many ways the government already regulates research where
that is necessary to attain some public policy goal. For
example, we regulate nuclear research to protect public
safety and the National defense; we regulate the use of
human subjects in research to ensure the dignity and safety
of people; we regulate laboratory standards both to protect
worker health and safety and the environment at large. (In
one sense, one might characterize the proposed regulation as
further worker health and safety and environmental regulation applicable to a restricted case.) All this regulation
--including proposed regulation of research using recombinant
DNA technology--does not intrude on academic freedom for it
does not attempt to control the thought processes of scientists. It merely limits the extent to which scientists may
exercise their independent judgment to put the rest of
society at risk.

That such regulation does not constitute an unwarranted intrusion into academic freedom does not necessarily lead to the conclusion that the concerns of scientists regarding inappropriate intrusions should be ignored. Certainly, such a regulation would risk impeding innovation through scientific discovery, and freewheeling innovation has been a hallmark of this Nation's success. However, the focus of the question should be on the impact of the regulation on innovation and the value of that innovation, and not on a concern for the ethics of invading the freedom of thought.

2. To what extent should academic research activities be distinguished from commercial activities?

To answer this question intelligently we must focus on attributes of the particular activity to be regulated--what kind of conduct occurs, what dangers does that conduct entail, and how can those dangers be controlled? Since the academic setting is different than the commercial, it seems likely that the answer to each of these questions will be different and, therefore, the regulation adopted will in some particularities be different. For example, the NIH Guidelines' restriction on large batch research may be fully appropriate for the academic context, but unnecessary for commercial laboratories with a history of skill in containing various forms of toxins. On the other hand, commercial researchers may seek to keep information from administrative agencies for commercial reasons, even though some of that information may be necessary to protect public health and safety. It would be inappropriate to assume that researchers work in academia apply some morally superior set of controls to their research than commercial researchers. However, it would be proper to recognize that incentives for those working in one field might induce dangerous practices which are unlikely to occur in the other and to tailor regulations to fit the different contexts. Frankly, it would seem that the regulation's administrator is likely to be better able to make those distinctions than Congress in enacting a statute, for the differences are likely to arise for a wide range of narrow points. Therefore, any bill that specifically directs the administrator to take account of those distinctions in setting or enforcing standards accommodates that need.

3. Should the regulation, if adopted, be enforced through self-regulation?

Using Self-enforcing bodies, such as institutional biohazard committees, has several advantages. It fosters independence; it focuses attention within the institution

on the need to comply with standards; and it provides a
local body which is readily at hand in the event of miscon-
duct. However, such arrangements contain the implicit risk
that the self-regulators--swayed by allegiance to their
institution--will not rigorously enforce adopted standards.
To forestall that possibility, it seems appropriate to
ensure that the law's administrator has the authority to
direct conduct of the private, self-regulatory bodies when
their conduct is not in accord with proper standards. Presumably, this would give the administrator authority to
require initiation of investigations, review disciplinary
actions, approve or modify institutional-specific rules, and
so forth.

4. In connection with adopting regulation of research using
    recombinant DNA technology, should Congress preempt
    State involvement?

I believe that Congress should limit its use of the
power to preempt State regulation to cases in which (1) some
important national purpose is served by preemption and
(2) the Congress is reasonably convinced that it can ensure
all the protection the public requires without State involvement. The national purpose served by preemption in this
case would be to ensure scientists working in any laboratory
in the United States would be subject to the same set of
regulations. It is difficult to see how recombinant DNA
related research deserves that protection from our federal
system of government more than any other research or
commercial activity. Further, preempting the field elimin-
ates the possibility that the States will experiment with
regulation, thus developing more effective models. Also, we
must remain aware that State agencies can assist federal
authorities in implementing regulations when a parallel State
and federal regulatory program exists. Thus, preemption has
the added disadvantage of eliminating this "branch office"
assistance at the State level.

5. Should we condition the grant of patents for discoveries
    resulting from research using recombinant DNA technology
    on a showing that the research was conducted in a manner
    consistent with the NIH Guildelines?

I think that we may both understate advantages and overlook disadvantages of this possibility. On the positive side,
imposing such a condition would present a tremendous incentive for investigators--and particularly investigators
looking to the commercial utility of their work to comply
with the NIH guidelines. Further, the regulation would constrain the manner in which foreign investigators engage in

research using recombinant DNA technology. While many of those investigators would not be subject to federal agency implementation of the NIH guidelines, they would be forced to comply with the guidelines if they are to market their discoveries under patent in the United States. This, in turn, has the added advantage of eliminating incentives to export domestic research.

On the minus side, disadvantages also loom large. The proposal is to adopt an enforcement device that is very inflexible. For example, it does not accommodate serenditous discoveries. We can anticipate that a patentable discovery using recombinant DNA technology will be made in circumstances in which the investigator had not intended nor anticipated being in the area of regulated activity. For example, as we know, research is presently underway to verify the standards set by the NIH guidelines. By definition, part of that research is outside those standards. The product of a serendipitous discovery during the course of that research might not be amenable to patent if the proposal were adopted, even though we might excuse an investigator from meeting regulatory standards because of the serendipity of the discovery.

This problem might be mitigated by making very definite and limited the area of conduct subject to the precondition for obtaining a patent. However, the more that definition is narrowed or made explicit, the more likely that conduct which would be subject to the precondition will be allowed to escape. Thus the likelihood that this type of regulation will not cover all conduct intended would increase.

Two other major disadvantages of this proposal also exist: (1) The proposal would not extend to all research using this technology, even in the United States. (2) The device necessitates an added layer of bureaucracy.

The precondition on patents will not cover all research for two reasons. Some researchers may be uninterested in the commercial value of their discoveries. Others may choose to avoid review for having followed the standards by relying on trade secrets, rather than patents, to ensure the commercial advantage of their discoveries. To avoid this disadvantage, one requires a more direct form of regulations--such as licenses, prohibitions, and government control of restriction enzymes--to ensure all domestic research is covered.

Further, there seems no way around the last disadvantage I perceive: The existence of the patent precondition requires an administrative staff to implement it. Perhaps

the matter could be simplified, for example, by requiring the
Secretary of Health, Education and Welfare to certify what
discoveries have been made in compliance with the NIH guidelines and to require patent applicants to obtain such certi-
fication as the condition to obtaining the patent. But,
some added cost and effort is involved.

An underlying principle also raises doubts about the
proposal. Indirect mechanisms complicate regulations. Preconditioning patent approval upon compliance with the guidelines indirectly does what can be achieved by licensing and
prohibiting in the United States and by treaty agreeing to
license and prohibit abroad. Such may be justified to
achieve goals not otherwise attainable. But, using more
direct solutions may help make compliance easier and more
predictable and may contribute to better relations with
other nations of the world.

On balance, I doubt the efficacy of preconditioning
patents on compliance with the NIH Guidelines as a useful
adjunct to more direct forms of regulation.

6. What information disclosed to the administrators of such
    regulation should be disclosed publicly?

Often the economic incentive to innovate comes from the
fact that the discoverer is able to keep the discovery
secret for a period of time, and thereby, enhance his share
of the market. Therefore, we are concerned with the possibility that regulation will force scientists--both commercial
and academic--to disclose the details of their discoveries
and lose those discoveries economic value. On the other
hand, we are all too aware that secrecy may breed abuse.
Therefore, we are concerned with regulations which are
shielded from the public eye. Balancing these worries is
no mean task. Personally, I believe that the balance proposed in Senate bill 1217 is judicious. That bill would protect generally the needs of secrecy but would provide for
the release of information--after proceedings to ensure that
all are afforded due process--where such release is in the
interests of public health and safety. The only modification that might be needed to that proposal is clarification
that the valuable secrets growing out of academic work
deserve the same protection as the secrets arising from
commercial research.

7. How should the products of research involving recombinant
    DNA technology be regulated?

One fascinating aspect of the debate over regulating

research using recombinant DNA technology is its focus on
research rather than the products of the research. How will
we decide what new discoveries should become publicly available? Current regulations cover some of those areas. For
example, using recombinant DNA organisms to manufacture
insulin which is ultimately sold to the public is no doubt
already subject to control under the Federal Food and Drug
Act. However, in other areas that is not so clear. For
example, what regulation is there to control decisions to
use recombinant DNA organisms to enable plants to absorb
nitrogen from the air? It is not surprising that we have
focused our attention on regulation of research rather than
research products since the area is so new. However, if
regulation is necessary, it would seem appropriate to include
in it provision to ensure that discoveries using recombinant
DNA organisms not be released into the environment without
some form of regulatory oversight.

II. A Menu of Perspectives

Permit me now to turn our attention to some theoretical
perspectives to provide the basis for a somewhat more comprehensive review of the way research using recombinant DNA
technology might be regulated. For all practical purposes,
no regulatory program successfully achieves the goals
spawning its enactment. Any regulation of research using
recombinant DNA technology which Congress may adopt has two
competing goals: (1) To encourage free and active research
and (2) to eliminate risks of danger to workers and the
public from such research. A new regulation with those
goals is likewise unlikely to be completely successful.
That, however, does not dictate that Congress, and this Subcommittee, should concede failure and forego efforts to
regulate. Rather, it suggests that the subcommittee judge
proposed regulations according to their propensity to
achieve legislative purposes and to their potential counterproductivity for such goals. Then the subcommittee must
compare the potential for success and counterproductivity of
proposed regulations with that of other proposals and the
status quo.

What are characteristics of a regulation most likely to
avoid dangers and least likely to discourage safe research
using recombinant DNA technology? Three principles may
assist finding the answer.

First, while regulations purport to dictate conduct,
they do not necessarily have that result. Indeed, sometimes
they dictate unintended conduct. For example, consider a
regulation which prohibits conduct. If it is not well known

or is well known but not enforced, it will not be very
successful at forestalling the prohibited conduct. On the
other hand, if enforced by imposition of severe penalties for
violation, it may not only preclude the prohibited conduct
but may also preclude related conduct. This will occur if
the regulated party fears that the enforcing authority will
extend the prohibition--and, therefore, the severe penalty--
to the related conduct. Thus, when prohibiting conduct,
Congress must consider (1) the tolerability of the unwanted
conduct's occurring occasionally, (2) the success with which
prohibited conduct can be distinguished from that not prohibited, and (3) the importance of having the nonprohibited
conduct continue. Depending upon those judgments, Congress
can select penalties of various severity. Also, it can
reduce the possibility of inadvertent violation. For
example, it can insert a licensing requirement to increase
awareness of the regulation among people involved in the
area of conduct. Or, under some circumstances, it can
restrict access to factors necessary to engage in the conduct thus disabling those prohibited from the conduct from
so engaging.

My second principle is this: Generally speaking, the
more flexible the tools available to the administrator, the
more likely he or she will be able to apply the regulation
in a manner consistent with Congressional purposes. By
flexible tools, I refer both to flexibility of standards and
of enforcement devices. An enacted standard is less flexible
than one adopted as an administrative rule, and that in turn
is less flexible than an administrative order. The law or
rule are more inflexible because they are adopted without
reference to their effect on each particular case; they
speak with a broad brush. Such standards likely have gaps
allowing conduct intended by Congress to be prohibited and
prohibiting that intended to be allowed.

Impediments to imposing more inflexible enforcement
devices, such as criminal laws, are less telling for those
more flexible, such as summary license suspension authority
or cease and desist order power. Hence, their application
is less likely to occur for several reasons. Criminal
prosecution connotes, socially, very significant wrong doing.
Thus, administrators quite properly in my opinion, often are
loathe to use that enforcement tool against people not con-
forming to regulatory standards but not appearing to be
truly "bad actors." Further, the presence in the enforcement
process of prosecutors, grand juries, judges, and juries in
addition to administrators increases the possibility that the
conduct, determined by the administrator to be violative,

ultimately will be ruled acceptable. Finally and related, the criminal justice system contains presumptions which preclude criminal conviction on facts sufficient for a cease and desist order, if permitted by the regulation.

Some considerations, however, limit the desirability of flexible regulations. An administrator with very flexible tools has broad power to apply the regulation either consistently or inconsistently with Congressional policy. While unlikely that administrators will be unwilling to comport with Congressional policy, they may be unable. When setting standards, they may not have the advantage of collegial debate or broad public interest which may arise by right of the matter's being in Congress. They may not have sufficient staff and budget to perform professionally assigned responsibility. Thus, the quality of their decisions may be poor.

Also, greater flexibility in a regulation may enhance uncertainty about acceptable conduct, and uncertainty can negatively affect innovation. That is the third principle I bring to your attention. We have already noted the "halo" effect that extreme penalties can have, discouraging conduct not meant to be prohibited. Uncertainty about standards, and the predictability of their enforcement, can have a similar effect. Indeed, it may be greater. The halo caused by extreme penalties only extended to activity not clearly distinguishable from prohibited conduct. Since uncertainty may spread over much more conduct than that likely to be prohibited, the halo will similarly extend. Uncertainty can discourage people for numerous reasons. Businesses will avoid incurring substantial R. & D. or capital expenses to enter a field from which they may ultimately be prohibited. Basic research investigators will not wish to commit their careers to research they ultimately will have to cease before completion.

Having already seen the case favoring flexible regulation, one cannot conclude, however, that regulations ought to be made inflexible in order to ensure certainty. A balance must be drawn. And, that balance must take into account the peculiarly difficult problem posed for regulation of research using recombinant DNA technology. We know very little about the actual dangers, although we guess that they might be great. Therefore, to set inflexible standards for conduct today would be folly. The probability is great that the standards set today based upon incomplete information gathered in the early part of what promises to be a long investigation--will be sometimes more and sometimes less stringent than ultimately appears appropriate.

III.   The Scenario of Permitting Some Research Using Recombinant DNA Technology

To allow research using recombinant DNA technology in order to realize foretold beneficial innovations but to limit it to avoid unacceptable worker and public health and safety risks requires a complicated pattern of regulation. Since the regulation should permit all research not taking "unacceptable risks," the initial question for designing a regulation is how to determine what risks are unacceptable. Such is not amenable to precise determination; Congress is left with two alternatives. First, it may proscribe identifiable conduct which has a high likelihood of involving the unacceptable risks. Prohibiting unqualified investigators, or investigators using unqualified facilities, from engaging in this research fits this alternative. Second, it may decide what risks are acceptable or identify an individual or group in whom it delegates that responsibility. Such an alternative ranges from Congress' relying on judgments of each qualified individual investigator working in his or her own laboratory, to its relying on some form of peer review and oversight authority, to its relying on decisions of a Federal Administrator, to its enacting standards, and combinations thereof. Both of these alternatives appear likely candidates for regulating the use of recombinant DNA technology.

Publicizing the prohibition of unqualified individuals --or of individuals working in unqualified laboratories-- from engaging in this research will deter much of the unwanted conduct. Likewise, providing both flexible and severe enforcement tools against violators will deter some or all intentional bad actors from violating the prohibition.

Nevertheless, the prohibition will be ineffective with respect to those who do not know about it and those not deterred by it. If Congress determines that a significant number may fit these categories, it may search for ways to deprive unqualified individuals and laboratories of the capability to undertake the research. For example, in the instant case, it may set limits on those who may manufacture, sell, buy, or possess restriction enzymes. Such a solution by itself would be insufficient to prohibit research by the unqualified since these enzymes can be manufactured in private laboratories. On the other hand, manufacture is difficult, and thus the more unqualified would be effectively disabled from engaging in this research.

The combination of two control techniques would work to reduce to a very small number the group of unqualified individuals in unqualified laboratories who might engage in this research. The number can be reduced even more by increasing the severity and variety of possible penalties imposed. Introducing penalties for the qualified manufacturer or seller who distributes to the unqualified will reduce the number yet further. Indeed, the outward limits of these control techniques can be very severe. However, if too severe, the "halo" effect, noted above, of precluding related but acceptable conduct will grow.

A simple prohibition of those unqualified from engaging in research using recombinant DNA technology will keep the vast bulk of people from this activity.

But, the problem becomes more difficult when drawing a line between those which the regulation consider qualified and those not quite. Administratively, it is easiest to enforce that distinction by requiring all who wish to engage in the research to obtain a license. The administrator can then review the qualifications and conduct of each.

Such a requirement, however, has the deleterious effect of increasing the cost of the research activity, because the investigators, including those unquestionably qualified, must obtain the license and suffer the delays and bureaucratic impositions attendant therewith. Further, given the research using recombinant DNA technology should only be conducted in facilities that are qualified for that purpose, it might be possible to substitute licensing laboratories for licensing individual investigators. In such a scenario, all investigators would have to be prohibited from engaging in this research outside licensed laboratories, but that seems to be a necessary standard in all events. Then, laboratories would be licensed and investigators would not be.

Such a plan has the advantage of reducing impositions on investigators. However, it poses the problem that unqualified investigators might be employed in licensed laboratories. Such difficulty might be ameliorated by authorizing the licensing agency to suspend or revoke laboratory licenses of laboratories employing unqualified investigators. Also the licensure system might borrow a concept from the Securities Exchange Act of 1934, giving the licensing agency some power over individual employees in licensed laboratories. For example, the agency might be given the authority to order a specific individual or group to cease and desist specific activity or to order that they be barred from employment by a licensed laboratory for a time specified or permanently.

Such enforcement tools have the added attraction of improving the administrator's effectiveness. Without them, he or she might be confined to revoking the license of a laboratory in which individuals are out of compliance, and in specific circumstances, that might be too great a penalty given the violation.

Licensing laboratories and not individuals is not without its disadvantages. Identifying the laboratory as an entity subject to regulation may not be simple, especially since most are not separate corporations. Ascertaining lines of responsibility among investigators within the laboratory may also be difficult. Other problems may exist. Presumably, when working out details of a proposal these can be overcome.

To my knowledge, one more significant problem with licensing remains: What conduct is to be made subject to the license requirement? At the risk of delving into a scientific question beyond my ken, I understand that research using the recombinant DNA technology has the following characteristics: (1) It is reasonably easy to distinguish from other research. (2) There are two categories of research using that technology: one justifies the regulation and the other does not. (3) A list of specific research activities can be drawn in which each item can be allocated to one or the other category. Finally, (4) such a list cannot feasibly be exhaustive.

Recalling the concern that the regulation not foster uncertainty, a twofold approach for identifying activity required to be licensed might work best: First, the regulation would list those research activities required to be licensed and those not. Second, it would provide a conceptual definition for those activities not required to be licensed. Such an approach allows investigators certainty about conduct not required to be licensed when that is possible, and allows independent exercise of judgment when the case cannot be predetermined. Variations can be selected to fine tune the balance between regulation and independence of the investigator. For example, one might give the administrator the power to develop the list activities not required to be licensed or to expand the list based on information learned after enactment.

Conclusion

In sum, I believe that we have a tendency to over-estimate the value of regulations enacted. Notwithstanding, I believe that regulations can be designed which reduce

risks of public and worker hazards significantly and which do
not unduly impinge on innovative research. To do so, however,
is a complex task requiring a thorough understanding of this
field of research of regulation of other activities, and
of the workings of administrative agencies.

Part IV
Regulations Protecting Human Subjects

14

Conditions and Consequences of Consent in Human Subject Research

Albert J. Reiss, Jr.

Introduction

The behavioral scientist's access to information is limited by important proprietary rights in information and by individual and collective rights to secrecy and privacy. Governments assert rights to keep information secret or confidential to protect national security and the deliberative and administrative processes of executive, legislative, and judicial agencies. Corporations and voluntary bodies, such as professions, have legally guaranteed proprietary rights to information to protect the autonomy of the organization and their client's right to privacy. Private persons, similarly, have proprietary interests and a right to security of private personal expression and affairs (1).

In a free and open society, these proprietary interests and private rights confront public rights and claims to information. What is available in the public interest depends upon both law and custom, including the customs of a scholarly community, and interpretations in any given case as to what is public and what is privileged. The federal Privacy Act and the Freedom of Information Act among others define rights and privileges in access to federal agency information.

The behavioral scientist's access to information is normatively a matter of right to information that is public and a matter of consent where it is proprietary, private, or privileged (2). How to regulate the acquisition, processing, and dissemination of public and private information is especially problematic in a free and open society. At the present time regulation of information is in a state of flux. While recent law makes information of public bodies more accessible to inquiry, that of private persons and organizations at the same time has become less accessible by government regulation.

Government regulation has arisen as a response to allegations and instances of malpractice and misfeasance in experimental research on human subjects. Control by scientific custom, professional ethics, and tort liability are deemed insufficient to protect participants in research from harm. These controls are largely supplanted by government regulation of research through the power of financial support. Our major concern here is with the social control of knowledge by government regulation, especially those issues and problems that arise from regulation based on a Bio-medical model of human subject research.

The Bio-Medical Human Subject Model

The major research model underlying federal regulation of behavioral science research was developed from bio-medical research. This research model assumes that a _principal investigator_ acts directly upon or intervenes in the life of private persons as _subjects_ of an _experiment_. In its elementary medical form, the subject and principal investigator have two role relationships, those of doctor and patient and of experimenter and subject. The basic regulatory model imposed upon this bio-medical human subject model is that investigators must not exercise any constraint or coercion while providing the essential information for the subject to make a free choice about participation in the experiment, a doctrine of _informed consent_. The information must include an assessment of the risk of harm to subjects and their potential benefits both to aid the subject in making a free choice and to assist regulators by using a _harm-benefits_ ratio to determine the propriety of sponsorship.

Useful as the bio-medical model may have been in articulating basic elements in the current regulatory mode, it has limited utility for other models and conditions more common to behavioral science inquiry. We shall discuss below the limits of this basic model for inquiry in some other models of research, principally those of research on organizations and by means of social observation or by modes of inquiry where there is no direct subject-observer contact.

Major Limitations of the Bio-Medical Human Subject Model. The parties and roles in behavioral science inquiry often do not conform to the elementary model of a one-to-one investigator and subject relationship. Much research is undertaken by a team or organization where a reasonably large number of employees as well as investigators acquire and have access to information regarded as confidential. The principal investigator often acquires none of the data, relying upon others to do so, and often operates primarily in the roles of administrator and principal analyst. The subject of inquiry moreover, frequently is an integral social group, organization, or collectivity rather than a person. Confidential information often is obtained by indirect rather than direct inquiry or from confidential records (3). Consent for access to confidential information may be sought from administrators of records or from parties other than those who are the original source of information. A growing number of studies depend upon systematic observation of natural social phenomena where the consent of the observed is not regarded as problematic. Visual and audio methods of acquiring and storing information and computer storage and processing both facilitate and complicate problems of identification and access to information.

The prototype for behavioral science research perhaps is the sample interview survey. In the sample interview survey, sampling statisticians select addresses where respondents are approached for consent to be interviewed by interviewers who are under no immediate supervision. The work product of interviewers is reviewed by a supervisor who may also make direct inquiry of the respondents to verify information and to audit interviewer conduct. This information in turn is transmitted to a field office where confidential information may be processed by coders and analysts <u>before</u> identification is removed. Still others will prepare the data for computation and analysis in a chain that ends with the preparation and dissemination of research reports. Some, if not all, of these specialists may need to have access to confidential information that identifies private parties. Few respondents in a survey who consent to participate by being interviewed could readily comprehend or become aware of this chain of accessibility to their confidence. That principal investigators can guarantee confidentiality under these circumstances is questionable. What is remarkable perhaps is how little evidence there is that such trust and confidence in interviewer employees is misused or broken.

Formal rules for certifying human subject research typically do not confront the question of who is qualified to secure consent from subjects. Generally the principal investigator's qualifications are taken as the criterion for approving the solicitation of consent. A similar situation prevails for approval by Institutional Review Boards where the reputation of field staffs, survey organizations, and other specialists in eliciting information is taken as evidence in approving the solicitation of consent.

It is no simple matter to control the activities of persons whose task it is to elicit information. Generally in
social research there are part-time as well as full-time
white-collar employees, who have been trained in a particular
eliciting procedure. Much social research is conducted by a
spatially dispersed set of employees who are not subject to
direct supervision and often are not under the direct control
of the principal investigator. Their competence will vary
considerably. This makes the fiduciary relationship between
investigator and task specialist and of the latter with the
subject precarious in two ways. First, the subject is vulnerable to incompetence and unauthorized misuse of information as well as fraud in failing to secure informed consent.
Second, the principal investigator is vulnerable to the employee's misuse of procedure and information, thereby increasing his legal liability while decreasing the integrity
of the research process.

Procedural competence can be partially controlled if the
principal investigator either monitors or seeks ways of determining employee competence to undertake the task of eliciting information. Yet in all employing organizations there
are failures, and a research organization is no more invul-
nerable to such failures than is any other organization. It
is in fact quite remarkable that misuse of confidentiality
and poor practice rarely cross the threshold to become regarded as problematic in behavioral science inquiry. At
the same time it must be said that very little attention has
been given to these matters. Where confidentiality is essential to the design of an investigation, principal investigators and Institutional Review Boards should seek information
on the competence of those who elicit information.

The subject participants in the bio-medical model are
institutionally organized and become available for research
because they are available through organizations. A bio-
medical subject is typically a client or patient who requires
treatment by professionals and para-professionals who practice in offices, clinics, or hospitals. These institutional
settings and their employees lie beyond the control not only
of the subjects but often that of the investigator as well.
The fact that many subjects become accessible to investigators because the investigator offers a professional service
within an organization is a critical element in regulation
of scientific inquiry. Many research subjects are unable
to distinguish among professional and scientific roles and
what is open to choice when dependent upon professional services.

Bio-medical interventions often pose a risk of physical

as well as psychological harm, a risk commonly lacking in
behavioral science inquiry. Most commonly, behavioral science inquiry seeks information by observation or interview.
There is seldom any harm in the procedure for acquiring information. Harm if it occurs, arises mainly after information is collected--from the public disclosure of private
matters that were communicated as a confidence. A fiducial
relationship is at stake. Once an investigator or his organization obtains information where disclosure can cause harm,
the investigator is potentially an agent for doing harm.

There is likely to be misunderstanding about what it is
that investigators agree to and can protect. Investigators
can protect only the information they acquire from their
being an agent of disclosure. They can offer reasonably adequate guarantees of that protection only when they have legal protection against its compulsory disclosure and strong
sanctions to protect it from illegal or unauthorized misuse.
Yet research participants and others are all too quick to
conclude that what is being guaranteed is protection from the
information becoming public knowledge, forgetting that the
only protection afforded is that the investigative agent
agrees not be an agent for its disclosure.

To understand their confusion, let us return to our elementary model of a single investigator with a single subject.
A guarantee of protection against disclosure of confidential
information is possible there only when an investigator acquires information uniquely through his intervention and it
is not communicated to the subject. Whenever a subject com-
municates knowledge to an investigator that already is known
to other persons, they also are potential agents of disclosure. Paradoxically, for the investigator, the subject is not
bound to keep matters confidential and may share the information with other persons. Probably in most situations where
confidence from public disclosure is promised by investigators, the guarantee can apply only to the agent as a source
of disclosure, not to disclosure per se, since disclosure by
other persons who have the information always is possible.

Whenever there is a third party to confidence--as is the
case when information is secured in group settings or of corporate actions--disclosure is possible without involving the
principal investigator as a source of harm. Unless third
parties who are present are bound to confidence, confidence
as such does not occur between subject and investigator.
Paradoxically, when there are multiple sources of disclosure,
investigators often may be perceived as a source of disclosure since they have few ways of demonstrating they have
not been agents of disclosure. The most that can be done in

such situations is to demonstrate that reasonable protections
against disclosure have been taken and that the investigator
has no compelling reason to have been the agent of disclosure. Given the fact that disclosure from more than one
source is possible in most behavioral science inquiry, it is
incumbent upon investigators to advise persons from whom potentially harmful information is sought that the protection
offered--if legal or other protections are afforded--pro-
vides no guarantee against disclosure by others. Similarly,
protection of subjects requires they be advised of any third
parties who may be privy to the confidence who are not bound
by agreement.

Regulations requiring informed consent from participants
in research are inextricably interwoven with regulations requiring the protection of confidential information for a number of reasons. First, there can be risks for parties who
refuse as well as those who agree to participate in a particular study. Should the investigator be legally compelled
to publicly disclose the fact of refusal or if a refusal
otherwise becomes public knowledge, the disclosure may damage the party who refused. Consider making public a list of
persons who refused to participate in a study of "former
patients in a drug addiction center," a study of "homosexual
networks," or a project studying "persons discharged by their
employer". Might not such disclosures cause considerable
damage to reputation and substantially risk future opportunities and benefits for those who refused? A particularly
thorny problem thus is raised about approaching persons for
their consent when even the knowledge of that approach is
potentially harmful.

Second, behavior science research occurs in diverse settings over which principal investigators and their agents
have little direct control and usually only limited indirect
control. Indeed, often they enter a private place where
others are present and the rounds of social life go on. As
a result of being admitted to private places or as an unintended consequence of a research procedure, information often
is acquired that was not intended as part of the designed
inquiry. That such information could be potentially harmful
to the person who granted consent for a particular study
is quite obvious. That the investigator often may not have
wanted to become party to the matter should be equally obvious.

Third, where confidentiality must be maintained to protect the parties from whom information is obtained, a requirement that one advise of the risks that might reasonably
be expected may prove at times unusually burdensome, given

the limited capacity of investigators to predict what information will be secured. Often one lacks sufficient knowledge about persons or organizations and what might prove damaging to them by disclosure. Although persons have a right to refuse information, if they have not done so, it may be a consequence of their difficulty to predict the consequences of that disclosure--particularly in the prototype interview situations for eliciting information in behavioral science inquiry. Investigators often lack information on how the information they seek might easily turn out to be harmful, since there is no established knowledge on such social harm and investigators are far from omnipotent. The inability to make such predictions precludes even a rough calculation of a harm/benefit ratio.

Moreover, there are types of behavioral science research where a harm/benefits ratio is inappropriate. The harm/benefit ratio is often inappropriate in the study of corporate actors. First, investigators often promise benefits from the results of an inquiry and the harm from disclosure of confidential information often is unpredictable. Second, in some cases, what is social harm may be simultaneously social benefit. A conclusion that a substantial proportion of banks have high risk investments can bring harm to these banks by bringing on an investigation of all banks during the course of which their condition is discovered and sanctions applied. At the same time, the disclosure may lead to increased control of the banking industry in the public interest--a rather clear social benefit. It should be apparent that this instance is rather different from the oft cited bio-medical example where one must first do harm to cause wellness, or to say that the first action is not harm since its intent is wellness. Social scientists may have similar examples but not infrequently the same information causes both corporate harm and corporate benefits, albeit it to different as well as to the same corporate actors.

The matter of protecting the integrity of corporate bodies is one that is particularly troublesome for behavioral scientists. On the whole, little attention is given to preserving the anonymity of private matters of corporate bodies. The socially harmful consequences of such disclosures are rarely examined. At times, the investigator may actually "intend" harm, as research undertaken in the spirit of muckraking sociology or social criticism (4). Social harm may flow also from the design of evaluation or action research studies where the disclosure of identity is built into the study design.

Risks of damage or harm exist as well for corporate

bodies that are the sponsors of behavioral science investigation. There is ample evidence of the political risks to institutions and organizations occasioned by scientific research (5) and behavioral science investigation (6). Behavioral scientists and their sponsors also assume political risks in competing with journalists (7), lawyers, and other organized modes of inquiry as they challenge more traditional and established modes of inquiry with claims of "scientific truth" (8). Congressional investigations of private foundation funding and of grants from public agencies for research into controversial social issues and of their ethical standards in research on human subjects impose political risks and governmental control over inquiry. On the whole, behavioral scientists have been given to view these investigations as attacks or threats to academic freedom and free inquiry. They are less commonly viewed as risks and moral dilemmas for sponsoring organizations, which they often are as well. The moral dilemma of university sponsors such as that of Harvard University faced with a broad mandate to protect sudents, academic freedom, and the reputation of the university in the psilocybin research of Leary and Alpert (9) is given much less attention. Yet in that case, as in many others, research sponsors are moved to institute controls over investigation as a resolution to political and moral dilemmas. The moral imperatives of protection with their attendant risks become a central focus of any organized effort to control bio-medical and behavioral scence inquiry.

The dilemma thus created for all parties to the research--principal investigators and their employees, research sponsors, and research participants--should be clear, but it is perhaps most critical for the informants or participants. Unable either to forecast what will be covered by the research design or to comprehend fully that which is and is not in a particular instance covered by the research mandate, the best advice one might give prospective participants perhaps is to refuse to participate if for any reason the participant expects that any confidential information will be secured that may be harmful. In any case, all parties should be aware of the fact that others who are not connected with the research process may decide what was not part of the inquiry and that all parties are unprotected in such matters. Indeed, third parties to an inquiry are rarely bound by confidentiality and consent. Without protection for confidential or private matters that are acquired apart from the intent of the research then, investigators not only must make judgments about the likelihood potentially damaging information might be acquired through their particular design but from the nature of their settings and the publication of their findings.

Finally, regulatory provisions that participants and investigators have evidence of informed consent poses special problems. First, informed consent that is written and documented by signature can serve to constrain refusals to respond or <u>withdraw</u> from participation. There is some evidence that signing any consent document makes it more difficult to break a trust relationship or agreement. The basic fiduciary element in any contract is not that easily broken--no matter how fragile it may seem in a modern world--and some participants will find it harder to break the relationship of commitment than others. It is well to remember that to bind investigators to do right may also bind their subjects so that they are less rather than more free. An informed and documented consent has such elements.

Subjects or participants, moreover, often are willing to consent but not <u>sign</u>. Signatures arouse suspicion and affect the willingness to participate. Whenever a unique identifier is not essential to the research study, it seems burdensome to require a signed or third party attestation procedure since on the average it will reduce the participation rate (a source of error) and may affect the validity and reliability of information (other sources of error). Above all, the requirement makes it more difficult to control and measure error in estimation for aggregates.

Second, documenting informed consent by participant signature (written consent) makes the unique identity of each participant known to an investigator, even when the object of all other procedures is to insure anonymity to all participants. Having information on the unique identity of participants creates ipso facto (ipso jure) a problem of protection where confidentiality characterizes the information.

It is unfortunately the case that modes of accessioning participants in behavioral science research often must provide for knowledge of one or more identifiers which can lead to unique identification. For example, if one wishes to have a random selection of households or persons in the United States, one must obtain information on certain identifiers such as an address and on household characteristics to select a respondent, e.g., "head of household." Since in a given unit at a given address, the head of household is often a unique identifier, e.g., in one person households, a single identifier can provide unique identification.

Yet it is well to bear in mind that in much behavioral science research, our interest does not lie in these identi-

fiers as a means of identifying unique individuals but in
social aggregates. The identifying information is incidental
to the participant accessioning or data collection procedure.
To return to our examples, we do not select an address or a
phone number by random means to know whom we are uniquely
getting information from but to insure that in the aggregate
we are getting information from participants who represent
classes of participants or who in the aggregate will describe
the universe of participants in which we are interested within a given range or error of estimation.

Although procedures then may make it necessary to col-
lect information that falls in the class of identifiers that
individually or collectively may lead to unique identification, any procedure of securing consent should not invariably coerce the collection of a unique identifier such as a
signature. Whenever identifiers are not essential to the
study design, documentary evidence of unique identity is burdensome since it increases risk of disclosure and correlatively the need for protection of confidentiality. That
risk of course can be balanced by forms of legal protection,
but the only certain way to protect is not to document accessioning participants and their informed consent with
unique identification procedures.

Competing Research Models

The Bio-Medical Human Subject Model is but one of a number of models to be considered for regulation of research
inquiry. Below we give special attention to two of them,
where organizations are actors and where information is acquired by direct observation of natural social phenomena.

Organizational Actor Model. Regulations for behavioral
science inquiry rarely regard organizations as actors. Yet
much behavioral inquiry is of organizations as actors, the
matter of principal concern here. Special problems of regulation arise not only because organizations are actors but
because participants who are members of these organizations
have roles both as actors within the organization and as private persons.

The granting of consent for intrusion into private
matters is a central issue when organizations are actors
since private matters may be those of corporate bodies as
well as of individuals. Individuals generally have information about their own private affairs, those of others, and
those of corporate bodies where they hold membership. A
corporate body, similarly possesses information about the

private matters of individuals (employees or clients) of
other corporate bodies (by intelligence or transactions) and
of its own corporate affairs. Clearly one important issue
is what may each consent to or provide information about
without having secured the consent of others on whom they
provide information? Correlatively, can an investigator secure information from corporate actors when that information
pertains to the private affairs of others? Any attempt to
answer these questions discloses that there are important
issues for both individual and corporate actors.

The question of who may grant consent is particularly
troublesome when information about private affairs is secured by indirect inquiry, i.e., from others or from the records of corporate bodies. Even where a corporate body has
secured consent to disclose information for use by others,
the agreement is generally so vague or incomplete as to lack
the basic elements of informed consent (3). A simple agreement that the information will be used only for research or
later treatment, for example, lacks the basic elements of
informed consent. The absence of specific legal prohibitions against divulging information that identifies individuals or corporate bodies leads to much questionable use of
files and dossiers of corporate bodies.

One of the more difficult questions about consent for
access to information on private matters arises in securing
consent on the private matters of corporate bodies since organizations often have no clear procedures for granting consent to gain access to such information. Employees, moreover, may purport to give consent when they lack authority
to do so or they make disclosures inadvertently. Without
written authorization for access to specific information on
corporate bodies, the legitimacy of acquiring such information is highly questionable.

The regulation of securing information from corporate
bodies may be impossible to maintain when adhering to the
following principles for who may grant consent. First, information on the corporate actions of individuals shall be
obtained only by informed consent on <u>direct</u> inquiry from the
individual. Second, there shall be no <u>indirect</u> inquiry on
the corporate actions of others or access to such information
from corporate bodies when the individual or corporate body
can be identified by the investigators. Third, information
on the private affairs of corporate bodies that identifies
the body may be obtained only on written authorization of an
individual or group of that body that has authority to grant
consent on its behalf.

One need not reflect long to see that such principles may fly in the face of social reality. Many personal or seemingly private matters arise in interactions that inextricably involve the private affairs of all parties in the interaction. Questions of the husband about the marriage relationship usually disclose "private affairs" of the wife. Questions asked of children about their relationships with parents frequently pry into the private affairs of their parents. In general, private matters are by definition often personal matters since they inquire into what sociologists call interpersonal relationships. The same holds true for the relationships of corporate bodies with their clients and with other bodies.

Furthermore, on direct inquiry few persons or agents of organizations separate their personal view about others from disclosing facts about others. A research procedure, indeed, can capitalize on the fact that informants do not make such separations. The willingness of persons to disclose information about others often is used to reduce the cost of collecting information. In any case, what these and many other examples illustrate is that in the course of social inquiry, one simply cannot avoid acquiring information that could bring harm to others whose consent was not obtained. Psychiatrists are altogether familiar with this problem in treating patients; social scientists are altogether familiar with it in studying most aspects of social life.

Clearly in these instances the problem arises as to how to protect information from disclosure when the only alternative is to foreclose the possibility of inquiry altogether. For whole classes of problematic aspects of social life that involve the study of relationships or interrelationships among individuals and corporate actors and for certain kinds of techniques such as sociometric and social network analysis that are based on social exchanges or relationships, it is impossible to conduct inquiry without acquiring information on more than the party whose consent was obtained. There is no simple answer to these questions. The suggestion that the consent of all parties be obtained before that of a single party is obtained often is unworkable for in many cases the other parties are not known in advance. One cannot study friendship networks except by first discovering the friendship network. Suppose that some friendship networks include participants in a form of deviant behavior, e.g., homosexual conduct. If one began by delineating the network and followed this with queries to learn what it is that formed the basis of friendship only to learn then that it is a form of sex relationship, one is immediately privy to information on all parties to the network. Similarly, examin-

ation of transactions of a corporate actor or of its agents
may disclose collusory agreement, implicating all actors.
What seems required for the study of "private" relationships
and exchanges, whether among individuals or corporate actors,
is adequate protection from coerced disclosure of information.

Within hierarchical organizations, the question of who
may grant consent for what information to whom is particularly troublesome. Where such responsibility is officially
delegated within the organization, the problem may be resolved when the object of inquiry is organizational behavior
within the information systems of the organization. Yet
special problems arise when the information must be secured
from individual members of the organization. Is their consent also required?

Perhaps one needs to make a distinction between private
and public corporations and their employee rights to consent.
Where a public or government organization is the object of
inquiry, individual consent may not be required so long as
it does not intrude into private matters. One has a right
to information about public organizations and the behavior
of their officials so long as organizational rights are not
infringed. The matter is obviously more complex for employees of private corporations or bodies where many of these
matters are contractual as well. In any case, there is a
broader issue so far as public organizations are concerned.
Under what circumstances may government actors refuse to
participate in inquiry? This is no simple matter.

Among the special problems one encounters is whether
organizational consent is required when a public organization is the object of inquiry but the information on the
organization is secured solely from direct inquiry or observation of members of that organization or its clients.
Should one, for example, require the consent of teachers to
test the learning increment of students in their classes or
that of students, when teacher as well as student performance
is being evaluated? Correlatively, if delegated officers
refuse permission to study organizational actors but the
consent of members of the organization is secured, is the
consent of all parties essential? There are no simple answers to these questions, but regulators must bear in mind
that when there is substantial power to block the objectives
of inquiry solely because persons are given power to refuse
by virtue of their organizational power and position, then
consent should not always be required.

<u>Systematic Social Observation of Natural Social</u>

Phenomena. The systematic social observation survey of natural social phenomena is analagous to the interview survey except that direct or observer observation of naturally occurring social events or behavior is substituted for self-observations reported to interviewers (10). Observers typically move to the social settings of participants and must accommodate to their rules. The setting is ordinarily beyond any control by the investigator or the observers and largely subject to control by the behavior of participants or corporate actors. The investigator is there as a matter of privilege in private places and as a matter of right in public places. There ordinarily is no prior relationship with participants before observation and none is ordinarily expected on its completion. Commonly there is little contact and no communication between observer and observed. Persons observed usually are anonymous to their observers.

Most models assume there are actors who are at risk in the research situation and whose consent is required, a set of conditions that is unlikely to characterize social observation. Whatever risks there are in the situation, they are ordinarily those of everyday life and when they are not, the investigator is there as a matter of right and can observe whether in a research role or not, the ordinary conditions for consent to participate are lacking. One may even raise the question as to when persons are not participants in research. This is not easily resolved in systematic social observation of natural social phenomena in public settings. Apart from procedural difficulties in securing consent from persons who are behaving in "everyday life," persons observed in their natural settings are not recruited as participants to a research project.

Though a particular research design may call for systematic observation of only some and not other persons in social encounters, behavior usually is naturally organized in ways that preclude selective observation. Whether or not observers record all that is observed, they are privy to much that occurs. One cannot, for example, observe the behavior of the police without observing the behavior of citizens in their encounters. A requirement that the consent of the citizen be secured before one could study the behavior of the police in that interaction is not only burdensome but might well endanger both the police and citizens were it necessary to secure the consent of the citizen before the police could intervene. Indeed, the most likely result of a requirement that the consent of all parties be secured would be to foreclose observation altogether. This example points up a complication of studying behavior in natural social settings. Intervention to secure an informed consent can itself fundamentally alter natural social situations and the risks attached to them.

In much social observation research, the only risk that ever exists is a risk arising from the failure of the society to grant legal protection for information. Thus in many cases the question might shift to one of when legal protection should be given rather than whether protection is afforded by informed consent. Informed consent is crucial when something can happen to the person because of what the procedure of inquiry does directly to the participant, a condition generally lacking in social observation. Consent seems far less critical as a condition when the only harm that can occur arises from the disclosure of information-- a problem that is largely obviated by legal protection.

Implications for Regulation

Considering the diversity among research models, it seems apparent that regulation must take into account the particular research model guiding the investigation, e.g., whether organizations or persons are the actors. Were one to generalize across models, regulation could pay particular attention to the following elements: (1) whether or not the investigator-subject relationship grows out of a prior or a continuing relationship; (2) the balance of power between subject and investigator, ranging from subject dependent to investigator dependent; (3) whether the research setting is subject to control by either of the parties to the inquiry or by other parties who may create an imbalance in investigator-subject power; and (4) whether the procedure of investigation alters the condition of the individual or corporate actor. Rules, of the following sort might guide decisions. Where subject power relative to investigator power is low, the free choice of subjects should be guaranteed by particular means. Correlatively, where investigator power is low relative to subject power, the requirement of informed consent may be waived or abbreviated forms accepted. Another rule might be that absent any procedural intervention that increases the risk of everyday life, regulation is unnecessary if confidences can be legally protected. Unfortunately little is known about the risks of everyday life and to know more is itself an empirical question that involves research on human subjects and their organizations. There is some danger that regulatory agents will "create" risks rather than rely upon knowledge of risks. The substitution of informed guesses is hardly a solution to a problem of risk assessment. Yet, another rule might be that where the primary risk to individual or organizational actors is from public disclosure of confidential or uniquely ident-

ifiable information, information be provided on procedures
for storing, analyzing, and reporting information that is
to be gathered.

Generally, social scientists are interested in analyzing
and reporting data for large aggregates in which it is not
possible to identify individuals. It should be sufficient,
in many instances, simply to inform those whose consent is
being sought that one is doing a statistical study where
aggregate reporting makes it impossible to identify them with
any of the information that becomes public knowledge. Analyzing and reporting information for social aggregates or
collectivities is an important way of preventing disclosure
of uniquely identifiable information. When, for any reason,
a procedure of analysis or reporting data is to be followed
where it may be possible to make inferences about individual
identities, persons should be apprised that procedure is to
be followed. A statement, for instance, that the information
is to be presented as a case study and whether or how identity is to be cloaked in reporting is a minimum of what must
be communicated in such instances.

Protection of Subjects or Investigators? Though empirical proof is generally lacking, it appears that current
procedures and regulations to protect human subjects are
designed more to protect investigators than to protect subjects. This is so for a number of reasons.

First, one of the core elements of current regulations
is the written informed consent signed by the subject and it
appears to be more an investigator than a subject protection.
Subjects have little recourse to claims against investigators
once consent is given, unless under the unlikely condition,
there is proof of fraud or gross negligence by investigators.

A second reason is that while investigators are to protect confidential information by prior agreement with subjects, subjects are largely unprotected because by law investigators have no testimonial privilege from subpoena.

A third major reason is that subjects must have information from investigators to make a free and informed choice
about risks of harm and any benefits, but the operating rule
is that only in the aggregate must risks of harm be over-
ridden by risks of benefit. Aggregate risks may assist
actors to choose but they provide no guarantees for actors.
Aggregate risks, however, favor investigators not only because ordinarily investigators need only justify a harm-
benefits ratio for an aggregate but because the choice of

aggregate risk is under investigator guidance or fellow professional control.

Finally, a fourth major reason is that regulation tends to be dominated by scientific professionals in decentralized units representing their interests and those of the sponsors of research. Local institutional review boards, for example, are appointed by and represent the organization which has legal liability for principal investigators. The board is a producer rather than a consumer of knowledge based organization.

Although on balance current regulations favor protection for investigators more than subjects, they also fail to protect either one in important ways. A major problem is that both participants and investigators are vulnerable and unprotected in matters of confidentiality of information.

Considerable attention, for these reasons, needs to be given to alternative regulatory modes, particularly ones that afford maximum protection to subjects or to investigators if these are the objectives or that optimize for both, if that is the objective. We shall not spend a great deal of time on these alternatives here, except to call attention to two major strategies for regulatory models.

Alternative Regulation: Legal Protection. The first strategy involves legal protection. One major form of legal protection is a statutory privilege from compulsory legal processes of disclosure (11). These statutes--so-called "shield-laws"--are designed to protect investigators from legislative, executive, and judicial compulsory process. The second form is that of a confidentiality certificate (12). The confidentiality certificate protects from compulsory processes all identifiable information of individuals that is gathered in a particular inquiry, sponsored and funded by the particular federal agency issuing the certificate. This form of protection meets the needs of a particular research investigation or sponsor, but it leaves unprotected all research where the government is not directly implicated as sponsor.

The statutory privilege for investigators obviously has more far-reaching implications for the development of subject and investigator protection against disclosure than does the confidentiality certificate. It seems particularly important and critical in affording maximum protection from investigator disclosure to both subjects and investigators since the major source of harm in much social science inquiry

comes from public disclosure of information. There are other
important reasons why the statutory privilege is preferable
to the Confidentiality Certificate. The power to issue Confidentiality Certificates is given to a federal agent or his
deputies. They also have the power to approve any proposed
changes in them, once granted. These are enormous powers,
since they carry with them the power to disapprove and control changes in scientific inquiry. Currently, moreover,
there are virtually no guidelines in any regulations that
constrain the discretionary power of agents who issue certificates. Under current regulations, the power to issue a
notice of cancellation of the Confidentiality Certificate at
any time can pose serious jeopardy not only to the project
but to continuing relationships between investigators and
subjects.

Alternative Regulations: Investigator Warrants. Current regulations place the burden on participants in research
to assume liability to a substantial degree by providing for
their contractual written consent to be a subject of investigation. An alternative is for the investigator or/and his
sponsoring organization to indemnify the participant or to
warrant the conditions of participation. Among a number
of warrants or indemnities that might be developed for social
research, one is especially noteworthy, the investigator's
signed testimonial with, or without, a notice or provision
for complaint.

Both participants and investigators require some form of
protection in behavioral science research. In the matter of
informed consent, participants must be informed particularly
of risks and/or benefits so that they may make a free choice
about participation while investigators must be protected
from being compelled to be agents of harm toward subjects as
a consequence of their participation in the research. The
consent form should do both these things, but on balance insure the rights of participants more than those of investigators should choice among them be necessary. A procedure
which more nearly balances both participant and investigator
rights and their protection is the investigator's written
testimonial or warrant of informed consent. Under this procedure, the investigator, or his/her agent, warrants that a
person or corporate body, or their representatives, have been
advised of the required elements in securing their informed
consent and it was granted.

Although one may regard testimonials or warrant forms as
providing sufficient information to constitute a legal document for formal litigation, provision may and perhaps should,
be made to provide for less formal modes of adjudication of

complaint. Where there is reasonable risk of harm, it seems essential to adopt a complaint procedure regardless of the procedure for securing informed consent. Written notice of when and where complaint can be lodged should be a minimum requirement whenever informed consent must be secured. The reason for this seems obvious enough. Most participants either err in acquiring information about whom they are dealing with in the consent procedure or they fail to retain or recall the requisite information essential to lodging a complaint. The obligation to provide information documenting how and where complaint can be lodged should fall in any case on principal investigators and their sponsors.

If only a complaint form is required, provision should be made to advise a participant that on request information will be provided on the name and address of any persons with whom he had contact, including any who performed any procedure connected with a research intervention, any who had authorized access to any confidential information pertaining thereto, or any of unauthorized access, if known. There likewise should be an obligation, on specific request, to identify a government sponsor.

The Social Control of Knowledge

To this point we have explored some regulatory issues and problems that emerge from what Edward Shils calls the confrontation of autonomy and privacy by a free intellectual curiosity (13). Regulation of the acquisition of scientific knowledge falls to a number of major institutionally organized agents of social control: ethical codes and their administration by professional peers and associations; legal codes and their administration by executive and jucidial agencies: government regulations and their administration by government and research organizations; or, by peer review organizations. Our purpose here is to highlight a number of issues that arise from government regulation, bureaucratic organization of regulation, and professional or peer implementation of regulations, rather than to systematically compare the various forms for the social control of knowledge.

The Role of Government Sponsor and Regulation.
To a growing degree, governments have become the sponsor of all scientific inquiry. The current government models for regulating scientific inquiry on their face place the federal government in several protective roles: (1) protecting the government's general interest in the public's right to information and its particular interest in deriving specific benefits for its many functions (legislative, executive, and

judicial) by setting program standards and objectives for
research to qualify for funding; (2) protecting the rights of
investigators from too much government interference by providing for institutional and peer review and making public
the grounds on which applications are denied by the government agency; (3) protecting the rights of participants in research by establishing regulations requiring investigators to
secure informed consent of subjects, provide benefits that
outweigh harms and other considerations as a condition of
their sponsorship.

There are other ways that government research sponsors
assume the legitimate mantle of protector, but the right to
protect carries with it more than a responsibility to see
that protection is adequate and in the public interest. That
"something more" is a responsibility for the consequences of
their decisions and actions stemming from their role as both
sponsor and regulator of behavioral science inquiry. The
government wants "cures" to physical, psychological and social ills and it shapes its programs and funding to do research on them—a proper role, to be sure. Yet the more government by its policies and programs provides inducements
that shape what investigators do, the more it must pay attention to its responsibility for the consequences of that
research activity.

The more a government induces research that requires
experiment and evaluation or interventions in social life,
other than the interventions required by research procedures,
per se, the more likely it is to do harm as well as good.
This is so if for no other reason than that even with a low
probability of harm, the more of that kind of research, the
more harm that is done by research. Since risks from some
kinds of research are greater than others, however, the more
the government induces investigators into high participant
risk research, the more burden it should also assume for
failures and liabilities. Such burdens should not fall exclusively on investigators and their institutional sponsors.
If the government wants a cure to drug use and encourages research on drug use in human subjects, it has not only a
strong obligation to protect those subjects, but it should
incur some of the liabilities that may result from any harm
done. Moreover, it should not readily do harm by disclosing
confidential information or other means without overriding
public interest. All parties to government sponsored research, including the government as sponsor, must come to
recognize there can be affirmative responsibilities as well
as liabilities when harm is done, a responsibility to help
that may override liabilities that might otherwise obtain and
a responsibility to share in the costs from tort actions or

other forms of settlement. It seems insufficient to encourage protection by "informed consent," leaving the risks to
fall to those who consent and the liabilities to those who
are the immediate principals in securing it.

Two developments raise the spectre of comprehensive federal control of scientific inquiry into human affairs. The
first is that some Federal agencies, such as the U.S. Department of Justice and its Law Enforcement Assistance Administration, base their regulation on a centralized agency review
rather than a mixed model of peer and government review.
There is evidence, moreover, that local institutional review
boards are regarded as advisory to the Federal agency sponsor and its review authority. The second development is a
stipulation by the Department of Health, Education, and Welfare that all organizations receiving funding and having established local institutional review boards must review all
research on human subjects, regardless of its funding source
and whether or not it is funded. Compliance with this regulation opens the research community to control of all inquiry by federal standards and procedures and places the
autonomous investigator in the hands of his peers and sponsoring organization.

Just how pervasive Federal government control of any and
all behavioral scientific inquiry will become is as yet un-
clear. There is little evidence, thus far, that regulation
of human subjects has led federal sponsors of research to
direct political intervention to control inquiry. There is
some evidence of indirect effects, however. Some investigators report they have been denied access to conduct inquiry into controversial issues, particularly access to
school settings, on grounds the investigation intrudes unnecessarily into private matters. While adherence to regulation is in itself a restriction on freedom to inquire,
denial of inquiry on grounds it invades the privacy of others
can readily cloak both sponsoring motives. Indeed, it appears that the powerful rather than the weak deny access and
invoke privacy claims in the face of direct investigation,
whether by journalists or social scientists. No matter what
form control takes, however, sponsor regulation of inquiry
fundamentally changes the relationship of the investigator to
sources of information. This can involve certain diseconom-
ies for individual investigators (14).

Bureaucratization of Research Regulation. Regulation
invokes rule-making and decision by rules is the hallmark of
bureaucracy. Whether or not regulation of research on human
subjects leads to an increase in the scale of organization,

review and approval procedures lead to bureaucratic procedures for making and implementing decisions. These procedures
can easily prove burdensome and give rise to patterned eva-
sion within research organizations.

One can readily envision the growth of an elaborate
appellate bureaucracy to mediate claims by investigators that
they have been unduly restricted in their inquiry and by subjects that they have been improperly dealt with. One may
also expect that informal patterns of settlement will emerge
to circumvent that system. Governments, moreover, may be
challenged on the propriety of their regulation and the stif-
ling of inquiry. Restricting the role of government to that
of a regulatory agency which protects the rights of parties
to an inquiry perhaps makes it more of a disinterested agency
than occurs when at the same time it regulates subject matter of scientific investigations and fiscal decisions for its
support. In the long run such functions may well be separated in the interest of disinterested regulation.

Summary

This paper has treated matters of government regulation
of behavioral science inquiry. We have emphasized that informed consent is inextricably bound up in behavioral science
inquiry with the risks that attend disclosure of confidential
information. We likewise have emphasized that an elementary
Bio-Medical Human Subjects model of scientific inquiry is
often inapplicable when applied to behavioral science inquiry.

Along the way, we have tried to maintain that some elements are more or less distinctive of behavioral science
inquiry and how regulations must treat them separately.

Behavioral scientists are interested in aggregative data
for individuals, whether individual or corporate actors, not
in individual level data. Exceptions arise for evaluation or
assessment research, and their requirements may be different.

Behavioral scientists generally intervene in the life of
participants only to acquire information from and about them;
it is much less common that some form of intervention other
than the research procedure is undertaken. Where it does
occur, such as in experiments with human subjects and their
collective life, separate consideration should be given to
the problems that arise when a research role intersects with
an intervention role and to the consequences for research of
deliberate intervention for purposes other than research.

Behavioral science inquiry is generally low risk inquiry so that for much of it, requirements of informed consent seem unnecessary and burdensome. The main risk from
harm in behavioral science inquiry arises solely from the
disclosure of confidential information, the disclosure being
the source of harm. There is an obligation to protect participants from that risk of harm by disclosure, one that can
be obviated by a legal privilege against compelled disclosure and by legal penalties for unauthorized disclosure,
misuse, or illegal use.

Finally, we make note of the fact that legal regulation
carries with it its own consequences so that if regulation is
to be enlightened and in keeping with constitutional imperatives these must be investigated by behavioral science inquiry. Regulatory constraints should be as few as possible
when studying the effects of regulation on free scientific
inquiry.

References and Notes

1. Samuel D. Warren and Louis D. Brandeis, <u>Harv. Law Rev.</u>
   4, 193-220 (1890); Roscoe Pound, <u>Harv. Law Rev.</u> 28,
   343-65 (1915).
2. There are some statutory limitations on consent where
   proprietary interests prevail or when exchanges are
   privileged.
3. Abraham S. Goldstein in Stanton F. Wheeler, Ed., <u>On
   Record:Files and Dossiers in American Life</u> (Russell
   Sage Foundation, New York, 1969), pp. 417-37.
4. Gary Marx, <u>Muckraking Sociology</u> (Transaction Books,
   New Brunswick, N.J., 1973).
5. Edward Shils, <u>The Torment of Secrecy</u> (The Free Press,
   Glencoe, Ill., 1956).
6. Gideon Sjoberg, <u>Ethics, Politics and Social Research</u>
   (Schenkman, Cambridge, Mass., 1967), pp. 141-61.
7. I. L. Horowitz and Lee Rainwater, <u>Transaction</u> 7,
   5-8 (1970).
8. Note that it is not argued that behavioral scientists
   have a more legitimate claim to truth, whether or not
   it is made in the name of scientific inquiry, but simply
   that their claim to science opens them to political
   challenge.
9. J. Kenneth Benson and J. O. Smith in Gideon Sjoberg,
   Ed., <u>Ethics, Politics and Social Research</u> (Schenkman,
   Cambridge, Mass., 1967), pp. 115-40.

10. Albert J. Reiss, Jr. in H. Wallace Sinaiko and Laurie A. Broedling, Eds., Perspectives on Attitude Assessment; Surveys and Their Alternatives (Pendleton, Champaign, Ill., 1976), pp. 123-41.

11. Paul Nejelski and L.M. Lerman, Wis. Law Rev. 1971, 1085-1148 (1971); Paul Nejelski and Howard Peyser in The Committee on Federal Agency Evaluation Research, Protecting Individual Privacy in Evaluation Research (National Academy of Sciences, Washington, D. C., 1975), pp. B-1-86.

12. United States Department of Justice, Law Enforcement Assistance Adm., Federal Register 40, No. 186:44034-37 (1975); USPHS, Federal Resister 40, No. 234:56692-95 (1975).

13. Edward Shils in Daniel Lerner, Ed., The Human Meaning of the Social Sciences (The World Book Pub. Col, Cleveland, Ohio, 1959), p. 121.

14. Albert D. Biderman and Elizabeth Crawford, The Political Economics of Social Research (Bureau of Social Science Research, Washington, D. C., 1968).

15

Privacy Regulations
and Longitudinal Studies

Lee N. Robins

Access to records is vital to the accomplishment of follow-up studies. Researchers' access to appropriate records
has been extraordinarily limited since passage in 1974 of the
Privacy Act and the Family Education Rights Act. The Report
of the Privacy Protection Study Commission (Personal Privacy
in an Information Society, Washington, D.C., U.S. Government
Printing Office, July 1977) makes recommendations that would
improve this situation a good deal, but leaves a number of
unsolved problems. My goal today is to discuss how regulations based on the recommendations of the Commission might
be expected to affect the carrying out of any longitudinal
studies, and to make some suggestions for modification of
those recommendations. Before doing so, however, it might
be well to ask what follow-up studies are and whether they
are sufficiently important to warrant coping with the
special problems they present in maintaining confidentiality.

Special Features of Follow-up Studies

Follow-up study designs have in common that they link
individually identifiable data collected at one point in
time with individually identifiable data collected at another point in time. There are three major types of longitudinal studies: The first is the real-time prospective study,
in which cohorts are selected at Time 1 and information collected about them. After some elapsed interval, information
is again collected at Time 2. When a portion of the subjects
experience a planned intervention during the interval, the
real-time prospective study becomes a controlled experiment.
A second type of longitudinal design is known as the "follow-
back" study, in which information is first collected at
Time 2, and then related to information in existing records

This work was supported by U.S.P.H.S. grants DA 00013,
MH 18864, MH 14677, DA00259, and AA03539.

made at an earlier (Time 1) date. This is the design usual
in "case-control" studies. A third type is the "catch-up"
prospective study. This design saves the waiting time inherent in the real-time prospective study by selecting subjects
from <u>records</u> created in the past (Time 1) and immediately
following them up through records or personal contact at
Time 2.

Follow-up studies have become an important scientific
tool in studying a host of issues: normal human development,
the natural history of medical and psychiatric disorder, the
long-term effects of historical events ranging from natural
catastrophes such as earthquakes to technological developments such as television, and the success of governmental and
private efforts to produce change--whether through educational programs, social welfare programs, new laws, or treatment
of individuals for educational deficiencies, medical illness,
or psychiatric and behavioral abnormality. Not only have
follow-up studies been shown to be a useful research tool,
but for these purposes they seem to be an indispensable tool.
Designs relating outcomes to available population statistics
and cross-sectional designs have serious methodological drawbacks when dealing with these issues. Ecological fallacies,
retrospective falsification, confusion of the effects of
aging with historical changes, and errors in assigning temporal sequences are some of the issues that plague attempts
to substitute alternatives for follow-up studies.

The Contribution of Records
to Follow-up Studies

If we accept the need for follow-up studies, we must
consider what distinguishes those that produce valid results.
Trustworthy results from follow-up studies depend on being
able to select an unbiased sample of the population of interest and to obtain unbiased data about that sample for both
time periods. These goals can be achieved when three conditions are met: 1) the population rosters from which samples
are selected are complete; 2) the rosters contain enough
personal identifiers to allow locating almost all of the
sample members for personal examination or interview and to
allow definitively identifying their records in other rosters, and 3) assurances of confidentiality encourage both
sample members and record-keepers to hold back nothing relevant. For many follow-up studies, the researcher needs
access both to the subjects themselves and to records about
them. Seeing the subjects personally is valuable because
subjects can provide information that has never been recorded, and can also alert the researcher to records about
themselves that would otherwise be missed. Record searches
permit verifying the subjects' statements and also provide

information subjects cannot supply--either because their memories are faulty (e.g., they cannot accurately recall the order in which various life events occurred) or because they never had the relevant information (e.g., diagnoses physicians did not disclose to them). Records thus serve follow-up studies in a variety of essential ways: as population rosters from which to sample, as sources of identifiers to help in locating individuals for interview and for locating other records about them, as a test of the accuracy of interviews, and as primary data sources.

Can such follow-up studies be carried out successfully if access to records is not restricted by government regulation? Indeed, yes. Let me use my own experience as an example. Long before the enactment of the Privacy Act of 1974 and other federal regulations regarding confidentiality, I used records in a number of studies aimed at discovering childhood predictors of adult psychiatric status. These studies obtained unbiased samples of the population of interest by sampling from complete clinic records, school records, armed services records, and selective service rosters. By tracing addresses through many record sources, we produced extremely high follow-up rates after periods ranging from one to 30 years. In one 20-year follow-up, for instance, we located 98% of the sample and interviewed 95%. We also supplemented the interview with record information that both showed that the interviews were very candid (97% of heroin users by record reported their use in interview) and made it possible to determine unequivocally the sequence of events (for instance, how often high school dropout predated or followed first juvenile arrest) and to demonstrate the transmission of truancy from one generation to the next by comparing school absence records for parents and children. These records had the great advantage of being uncolored by teachers' biases, respondents' memory errors, or interviewers' sympathies.

In the course of carrying out these studies, we exploited our access to individually identifiable information about individuals in a great variety of record sources, perhaps in as many sources as any other study has. In a follow-up study of schoolboys into early manhood, for instance, we collected an average of seven different types of records per person. In addition, we collected some record data for their wives and children which, when linked to interviews and record information for the men, allowed us to study family units. Records used included birth, death, school records, marriage and divorce records, and police records for subjects, their wives and offspring. For the subjects themselves, we also collected records of correctional

institutions, voting records, drivers licenses, telephone
numbers, city directory information, public housing records,
county welfare information, records of all three military
branches, Veterans Administration claims records, unemployment compensation applications and receipts, psychiatric
hospital records, federal drug treatment records for pris-
oners, probation and parole records, records of alumni
offices, of credit bureaus, of outpatient clinics, of city
jails, and state and federal prisons, and even addresses
(although no income data) from the Internal Revenue Service.
We were allowed access to almost every record source we
approached, although we had no written permissions from the
subjects. The FBI turned us down, but we often found FBI
rap sheets in prison files. The Social Security Administration was legally allowed only to forward letters for us or
to provide earning records on aggregated subsamples. We
found substantial adult record information for almost every
man, enough to provide a remarkably complete picture of
their lives. Virtually everyone leaves a record trail, and
that trail has become even clearer since 1974 with the com-
puterization of many records.

Preserving Confidentiality

The fact that follow-up studies with high recovery
rates can be done when there is easy access to records says
nothing, of course, about whether they can be done ethically.
The chief danger, it seemed to us, was breach of confidentiality. In the course of our follow-up, we believe that no
respondent's confidentiality was breached. Among the rules
we observed to preserve confidentiality of records (and interviews) was never sharing any information we had with any
agency from which we accepted information. We searched files
ourselves whenever possible. We never hinted to friends or
relatives anything they were not certain to know already,
never mentioning time spent in jail or previous spouses or
illegitimate children whose names turned up at some point in
our record searches. And we never gave the subject reason
to believe we had any information other than what he chose
to tell us. To make certain that there could be no slip-up
in this separation of interview and record information, we
gave the interviewer only the name and most recent address,
without any information as to how we had obtained the
address.

Recommendations of the Privacy
Protection Commission

Since the Privacy Act of 1974 and the Family Educational Rights Act of 1974, access to this wealth of record

information for research purposes has been severely curtailed. The recent Report of The Privacy Protection Study Commission argues that current prohibitions in these two Acts unduly inhibit research. It underscores the important distinction between administrative uses as compared with statistical and research uses of records, and the necessity for different rules for the two. The Report recommends revision of the provisions of these Acts which make illegal the disclosure of records in individually identifiable form without written permission to exempt legitimate research uses of government, medical, social agency, and school records. With these changes it would again become possible to link records of some agencies, and to choose samples from these agencies' record rosters. While the recommendations of the Privacy Protection Study Commission go a long way toward making it again possible to do follow-up studies, they leave many of the problems special to longitudinal research unsolved. The Commission recognizes that longitudinal research is a specially difficult area because it requires preserving individual identities over time (p. 584). I would like to review those recommendations which most hamper the carrying out of longitudinal research and suggest revisions of them that I think would make follow-up studies feasible without creating excessive problems with respect to privacy.

Before doing so, let me point out that I am going to be discussing only the least problematic kind of longitudinal studies: the follow-up of relatively small samples chosen from large populations and studied by groups who have no administrative responsibilities for the subjects. When the research topic requires the study of small populations such as prominent people, organizations, or localities, or where the research is internal--i.e., carried out by an organization with administrative interest in its subjects, there are special problems in maintaining confidentiality. Conversion of data into statistical tables cannot be relied on to conceal identities when the universe of objects is small. When researcher and administrator belong to the same organization, there is serious risk of flow of research information to administrative uses. Of course, these are problems for nonlongitudinal research as well. I will also be assuming that no untoward invasion of subjects' privacy is involved in inviting them to participate in a research project, so long as they are free to accept or decline without prejudice. Sufficiently frequent invitations to participate in research projects could constitute a considerable annoyance, of course. Indeed, the Commission is concerned that students are an oversurveyed population (p. 420). This may often be true for patients in the teaching hospitals of research-

oriented medical schools as well, but given the current
level of research funding and the high cost of surveys in
less "captive" populations, there would not seem to be an excess of intrusions on the general population to invite their
participation in scientific studies.

In Chapter 15 on Research and Statistical Studies, and
in Chapter 7, which deals with the use of medical records
for research, provisions are made for enabling the research
access to certain records without first obtaining individual
permissions, but a number of recommendations still create
special problems for longitudinal research.

Access Only to Selected Records

The first issue I would like to raise is the Commission's selectivity with respect to which agency records are
to be available for research in individually identifiable
form without permissions. While the Commission provides
special encouragement to the use of records collected by the
federal government, schools, medical services, and social
agencies, it makes no universal recommendation about the
access of researchers to records of all kinds. The arguments supporting access to medical records by the Commission
are that 1) medical research is a major expenditure of the
government and foundations, 2) that the government is already involved in health care through Medicaid and Medicare,
3) that the government is likely to become more involved
through some form of national health insurance which will
require access to records to evaluate the quality of medical
care. The arguments for allowing the use of health records
for research without authorization are cogently presented on
p. 309. If studies were to be limited to those subjects for
whom permissions were obtained, the study might be invalidated by failure to locate or failure to persuade a proper
sample of such subjects to give permission; further, those
who were located and willing to authorize disclosure might
not be reached in time to answer a pressing public health
need, as in the example given of discovering the cause of
Legionnaire's disease.

These arguments apply equally well to records of other
kinds of agencies when one looks at health research in a
broad context. The government's interest in health encom-
passes not only disorders once they require medical care,
but the behaviors that are associated with subsequent mortality and morbidity. Life styles characterized by violence,
abuse of alcohol and drugs, traffic violations, unemployment

and absenteeism have been found to predict early death by
accident, suicide, and homicide; increased rates of
psychiatric hospitalization; increased use of emergency rooms
for injuries, acute intoxication, and overdoses; and a higher
rate of medical care of all kinds, because such individuals
have high rates of cirrhosis of the liver, cancer of the
lungs, mouth, and esophagus, peptic ulcers, alcoholic gas-
tritis, and serum hepatitis. Thus records relevant to health
are by no means limited to those maintained by providers of
medical services. Employment records, welfare records,
police records, credit ratings, military records--indeed the
full variety of records that I listed earlier as those we had
searched in our follow-up study--are useful in providing a
comprehensive picture of the kinds of life styles that for-
bode increased morbidity and mortality. We now know how
useful such records are only because they were once available
to us in individually identifiable form without requiring
permissions.

Records concerning employment, trouble with the law, and
trouble in the military are useful not only as predictors of
survival and health; they also allow measuring the effectiveness of medical and psychiatric care. Treatment for drug
abuse, for instance, is best evaluated by combining self-
report of drug use and symptoms with objective measures of
social functioning. Employability is also a crucial criterion for evaluating the success of treatment of chronic
physical disorders. The researcher's having to depend on
individual permissions for access to police and employment
records has the same drawbacks that the Commission recognized in exempting medical records from this requirement:
Since permissions will not be universally granted, there is
the danger that the assessments may be biased and not comparable across patients.

Rather than recognizing the possible utility of opening
such records that can serve as indicators of social functioning to research, the Commission applauds the Fair Credit
Reporting Act for requiring credit bureaus to disclose credit
information without individual permissions only for credit-
related purposes (Recommendation 13, p. 87). The Commission
recommends similar restrictions on employers' records
(Recommendation 33a and b, p. 272), limiting information that
can be disclosed without individual authorization to directory information (the fact of employment, dates, and job
title). Even dates of attendance at the job are to be revealed only to law enforcement authorities, presumably when
an alibi is in question. Thus work attendance, an excellent
measure of health, cannot be used for research purposes.

Expiration Dates of Permissions

Records other than those of governmental, medical,
school, and social agencies could, of course, be used for
longitudinal research if permissions were obtained. However,
the Commission recommends that authorizations to look at
agency records remain in effect for so brief a period that
longitudinal research would be seriously hampered. As an
example, authorizations given an employer to obtain records
about a potential or current employee should have an expiration date not to exceed one year (Recommendation 16g, p. 253),
and authorizations to insurance institutions should be
limited to one year, except in the case of life insurance,
for which the date might be as much as two years after the
date of the policy (Recommendation 8g, p. 197). Since
follow-up studies frequently need to follow their subjects
in records over a number of years, this would require yearly
reconfirmation of permissions granted, clearly an unnecessary
intrusion on privacy if subjects are willing to grant the
permissions for longer periods.

Limited Disclosure

Even with respect to records to which the Commission
specifically recommends access for researchers without individual permission, there are suggested limitations in their
use which make longitudinal studies difficult. One of these
limitations grows out of the principle of limited disclosure
of medical records (Recommendation 11, p. 313). The
Commission recommends that "any disclosure of medical record
information by a medical care provider, with or without the
authorization of the individual to whom it pertains, be
limited only to information necessary to accomplish the purpose for which the disclosure is made." Since it is not only
possible but probable that a follow-up study would not require all the detail in a medical record, this would seem to
make it necessary for the agency staff rather than the researcher to abstract records, to prevent the researcher's
accidentally seeing items of information that he had not requested. Such a requirement would both reduce the quality of
the data collected, since it would be abstracted by a clerk
without any special knowledge or interest in the needs of the
research group, and it would greatly reduce the opportunities
for reviewing records, since it would be costly from the
agency's point of view to comply with such requests.

Agency Judgments of Potential Benefits of the Research

The Commission recommends that the records of government agencies may be disclosed in individually identifiable

form only if the agency (Guideline 3c, p. 602)

> determines that the research or statistical
> purpose for which any disclosure is to be made
> is of sufficient social benefit to warrant the
> increase in the risks to the individual of
> exposure of the record or information.

This regulation seems to give the government agency which
holds the records the responsibility for making a judgment
about the scientific merit of the research plan, since poor
research is of no social benefit. This is a judgment that
most agencies are not equipped to make. Research projects
supported by government grants have already been through
peer review, the grant-awarding agency's council, and institutional human rights committee review. Aren't these sufficient hurdles to determine that the research is indeed worth
the risk? Assigning the responsibility for this judgment to
the agency whose records are involved might not be such a
serious problem if the research design involved only a single
agency. If permission were refused, the study would not be
done, but at least it would not be done badly. But if the
study intends to link records of multiple agencies, as
follow-up studies often do, the very arguments that the
Commission offered in favor of allowing the statistical use
of government records without prior consent of individual
subjects--that the requirement for consent would lead to
biased samples and poor science--applies equally well to by-
passing the individual agencies' judgment of the scientific
merits of the research. The failure of one agency among
many to agree on the usefulness of the study and its consequent refusal to disclose its records may result in biased
results and thus seriously compromise the quality of the
research.

The right to judge the scientific merit of the study
which the Commission gives the individual agency might well
be used as an excuse to refuse permission when what really
deters cooperation is a fear that the agency itself may be
harmed, should the study report inaccuracies in the agency's
records or other indications of agency failure. The
Commission Report is concerned only with the privacy of individuals about whom records are kept, not with "privacy"
needs of agencies themselves. On practical if not ethical
grounds, the Commission perhaps should consider requiring
that researchers not disclose unnecessarily the particular
agency that provided records nor gratuitously criticize the
functioning of cooperating agencies, when that agency's
functioning is not the topic of the research.

Restrictions on Redisclosure

Another troublesome recommendation is that no agency records given for a particular study may be used for any further research without again obtaining permission from the agency that initially provided them (Guideline 3e, p. 602). This recommendation would forbid a second round of follow-up without a second approval. For second follow-ups undertaken many years after the first, a reorganization of government agencies could make it difficult to decide who has the authority to give the requisite permission. Further, the identifiers originally provided by the agency are certain to have been amplified in the course of the first follow-up. Surely the identifiers developed by the research group are not the property of the agency providing the original identifiers. How can the original agency prevent their use in further research?

Destruction of Identifiers in Research Records

Once a research project has been completed, whether or not it is a longitudinal study, the Commission recommends that its records or information be kept in individually identifiable form only so long as is necessary to fulfill the statistical purpose for which the record or information was collected (Recommendation 4, p. 584). Destruction is also recommended with respect to records of unsuccessful school applicants (p. 434) after 18 months. The destruction of identifiers in research and school application records wipes out the opportunity to do any later long-term assessment of the subjects. For instance, with respect to school applicants, it might be worthwhile to learn whether denial of entry into a particular school has any long-term disadvantage to the individual in terms of his eventual level of education, his occupational status or any of a number of other possible outcomes. Meaningful results would not be obtainable until at least ten or fifteen years after his rejection--after time for a stable occupational status to have been achieved. Such an opportunity would be lost unless plans for the follow-up study were laid at the time of initial data collection. There are many interesting long-term follow-up studies that used records of studies with a much shorter initial design as their starting point--for instance, Vaillant's follow-up of the Grant Study, McCord's follow-up of the Cambridge-Somerville study, Block's follow-up of the Berkeley Growth Study. None of these was foreseen at the time of the "last" follow-up contact in the initial study.

Suggested Revisions of the
Commission's Recommendations

What changes in the Commission's recommendations could be made to improve the opportunity to carry out longitudinal studies without unduly increasing the risks of breaches of confidentiality in individual records?

The Report of the Privacy Protection Study Commission recommends opening government records not otherwise forbidden by law, medical records, educational, and social agency records to research in individually identifiable form. The Commission might well consider taking the recommendations it has made with respect to these records, and applying them to all record collections, both governmental and nongovernmental. They could recommend that, under appropriate regulations, as spelled out with respect to educational records (p. 441), government records now denied to research by regulations applying specifically to the Census, Social Security, and the Internal Revenue Service should become available for research. Similarly, they could recommend that all local governmental and non-governmental records be made available in identifiable form for research under similar regulations, thus endorsing the use of employment, credit, and police records for location and evaluation.

I would like to see the Privacy Protection Study Commission amend its recommendation with respect to the destruction of identifiers in research records after a fixed time period. Destruction seems an excessive reaction to the dangers of disclosure, given the value of possible later follow-up. If records can be protected for one year, they can certainly be protected for longer, and eventually, when the interested parties die, they do not need protection at all. The Census allows disclosure of identifiable records after 72 years and uses the National Archives as a repository for identifiers in the meantime. Whether or not this is the best solution for research records needs further discussion, but identifiers could be removed and deposited in some safe place, perhaps using one of the link-file systems devised by Boruch and others. The identifiers would then be ready for re-linking to the substantive data by future researchers. Safeguarding the link certainly is an important problem, but it seems to be a soluble one, while destruction of identifiers is an irreversible loss.

Records of school, medical, social, and government agencies would be more useful to longitudinal research if the recommended conditions for their use without individual permissions were modified somewhat. The current recommendations may be found under Recommendation 13 (p. 441) with respect to the use of educational records for research purposes. Items a, b, d and f are certainly acceptable safeguards. Item a, as I understand it, simply means that records may be used unless there was a promise of nondisclosure for research purposes at the time of data collection. (If that is not its meaning, then perhaps Item a also needs modification.) Item b states that disclosure be necessary for the research; Item d requires that the records be kept confidential; and Item f that no administrative actions be based on them. The remaining requirements, Items c and e, do pose difficulties. Item c requires that the agency that possesses the records must determine that the research or statistical purpose for which any use or disclosure is to be made warrants the risk to the individual from additional exposure of the records. I would like to see the Commission substitute a requirement that the research proposed has met tests for feasibility, scientific soundness, and ethicalness through peer review and institutional review processes. This would relieve the agency possessing the records of the need to judge the scientific merit of the research, returning that responsibility to the peer review system. Item e is the requirement that any further use or redisclosure of the record or information in individually identifiable form will require express authorization of the educational agency. In place of that requirement could be substituted the statement that any subsequent reuse of the data would be subject to all of the above restrictions. This would require any future follow-up by the same investigator or by transfer of the data to another investigator to meet all of the rules preserving the confidentiality of the data that were specified for the original study. This change, of course, is intimately associated with the change in Item c. If the agency itself does not undertake to evaluate the value of the initial research study, it has no interest in judging subsequent studies so long as its clients' confidentiality is protected. The changes suggested of course would change the meaning, but not the form, of what is now Item g, which says, "any disclosure must be pursuant to a written agreement with the proposed recipient which attests to all of the above agreements."

Protecting Confidentiality During Data Collection

The suggestions I have made thus far would reduce the severity of the restrictions on the use of records for follow-up studies, as compared with the recommendations that have been put forward by the Privacy Protection Commission. Both the Commission and I might feel more secure about

reducing these restrictions if at the same time, there were
spelled out in more detail the principles that researchers
must use for protecting the confidentiality of their subjects, and, where desired, the identity of the agency.
Among these principles might be included: 1) In case-control
studies, i.e., where affected and non-affected individuals
are identified at the time the study begins, the identity of
the individual as case or control must be separated from
other identifiers during the period of data collection, so
that there will be no disclosure of the case or control status to agencies in which information is being sought that
might have administrative interest in the subjects. 2) If
no subjects with "nonincriminating" identities are in the
study, lists of subjects must be liberally salted with random names so that agencies from which further data are
sought will have no assurance that any particular individual
on the list is of interest to them. 3) When information
other than identifiers is obtained from records prior to interviewing subjects, interviewers must be kept blind with
respect to that record information so that there will be no
danger of their disclosing such information to the subject
or others. 4) Research studies on sensitive topics should
be required to develop an innocuous description of the study
and innocuous letterheads for contacting employers or relatives for help in locating subjects so that mere participation as a study subject does not in itself give away any
information that interested parties might not already have.
5) To protect the agency itself, research groups might be
instructed to salt the lists of potential interviewees with
names from non-incriminating sources, such as telephone
directories, drivers' licenses, and city directories, so that
the interviewer is truly blind about which agency has provided a particular name, and so cannot tell the respondent
which agency provided his name. 6) Identifiers must be kept
separate from research data and both must be kept securely.

Legal Protection of Research Records

A subject's participation in longitudinal research in
most cases is limited to a review of records collected over
his lifetime and an interview. Participation in research of
this kind poses minimal risks for the subject other than a
breach of confidentiality. Methods like those just described
have been shown effective over many years for preserving the
confidentiality of data collected in longitudinal studies.
With these precautions about separation of identifiers from
data, and secure storage of both lists of identifiers and
research protocols, the greatest threat of breach of confidentiality appears to be the threat of subpoena. The Privacy
Protection Commission recommends protection from subpoena of

records unless they are being required for a very few special circumstances, including proof of the violation of privacy regulations by the researcher. Until such time as the Commission is effective in getting protection for research records, it would probably be wise to add a seventh item to the conditions under which identifiable records are given out to the researcher: that the researcher agrees to destroy identifying evidence regardless of the risk to his own liberty in case of a subpoena, unless that subpoena is in the interest of preserving a life or physical safety or is requested to facilitate the investigation of the researcher himself or his group.

Risk/Benefits for
the Record-Holding Agency

Finally, the Commission might consider whether it might not go farther in actively supporting the use of agency records for research than it has thus far. Reducing the legal barriers to access is helpful, but may not be enough. Administrators need encouragement and reassurance to be willing to use their freedom to give out records in identifiable form. In one study we conducted, we requested and received an opinion from the Department of Justice that there was no violation of confidentiality involved in a particular government agency's giving us names and addresses. When we met with the head of that agency after he received that legal opinion, he said, "I know I _can_ do it. But I _can_ go stand out in the middle of traffic too." Eventually, he decided to take a chance on us, and I think has not regretted the decision. But it is easy to understand his feelings. The administrator rarely sees a direct advantage for his agency in contributing records to longitudinal studies. And the costs in time, trouble, and risks are only too obvious to him. We need to find ways to improve the risk/benefit ratios for the agencies on which we depend for record information as well as for their clients.

16

Confidentiality
and the Independence
of Social Research

Eliot Freidson

A number of important issues are raised by the papers
read here today, but I would like to concentrate on what I
regard to be the most basic issue underlying all others--
namely, the relation between social research and the state.
Social research does not go on in a vacuum and is not
influenced solely by the institutional arrangements immediately surrounding data collection and analysis. It is
also influenced very strongly, even if often indirectly,
by its legal position in the state and by its relationship
to agents of the state. The very data it is able to obtain
hinge on that relationship. I wish to suggest here that
social research cannot maintain a position of independence
which gives it access to especially valuable and unique
data unless it espouses policies which are highly protective of confidentiality, even to some degree at the
expense of some data interesting to researchers.

The best springboard for elaborating this position is
Dr. Robins' excellent paper. In her paper, Dr. Robins
rightly points out how certain kinds of research--namely,
follow-back studies--are made difficult if not, in some
cases, impossible by present-day privacy regulations. She
recommends that access to individually identified administrative records be made easier for researchers so that
they will be better able to mount follow-back studies. And
she urges rescinding the Privacy Commission's recommendation that identifiers in stored research data be
destroyed so that more follow-back studies can employ such
data.

The grounds for Dr. Robins' recommendations are two-
fold: First, she correctly argues the value of the follow-
back design. Following individuals over time provides
more powerful sources of causal inference than cross-

sectional comparisons can provide. The follow-back method
is far cheaper than a prospective longitudinal design,
though as I shall argue, the data employed by the former is
more difficult to evaluate than data collected by
researchers from the individuals involved. Second,
Dr. Robins argues that researchers may be trusted to
preserve confidentiality when they have access to identifiers. I agree. Researchers are indeed honorable people,
or at least harmless, and may be trusted to make every
reasonable effort to protect those they study. They may
not all actually go to jail rather than divulge identities,
but I cannot hold this human fraility against them. I
shall argue, however, that this is not sufficient protection for respondents. Researchers must be prepared
to give up some of their research schemes if they really
wish to assure the protection of identities.

Surely anyone who has read any significant portion of
the report by the Privacy Commission will know that in the
case of credit, hospital, personnel, insurance, and a
number of public agency records, a very large variety of
people have had access to them in the past and are likely
to continue to have access to them in the future. In
arguing that more individually identifiable data be made
available to the researcher, Dr. Robins neglects to take
into account the fact that when identifiers are preserved,
they also become available to a large number of other
agents and agencies which have entirely different needs
and conceptions of propriety, and which can do serious harm
to those identified. While the researcher's access may be
harmless, therefore, others' access may not be harmless at
all. It is because individually identifiable records
become available to an enormous variety of individuals and
agencies, and not merely social researchers doing follow-
back studies, that I would urge destruction of identifiers
wherever possible--certainly in stored research data and
also in at least some kinds of administrative data.

It could be said that it is not necessary to destroy
identifiers because it is possible to create a security
system that would prevent unauthorized revelation of individual identities. Indeed, a number of sophisticated,
well-intentioned people have put their ingenious minds to
the task of inventing very elaborate security sytems to
protect individual identities connected with sensitive
research and administrative information. However, no matter
how elaborate such security systems, they cannot be sustained without legal support. No security system can
stand in the face of a court order or legislation authorizing

its penetration. Even testimonial privilege in the form of
state shield laws can promise only limited protection. It
is apparently assumed by those who place their faith in
security systems for the protection of individual identities
that legal support for such protection (or at least restraint in the exercise of their authority by agents of the
state) has been neither problematic in the past nor will be
in the future. This assumption is naive and wishful.

Even a cursory examination of Supreme Court decisions
over the past ten years shows the lack of any strong and
consistent concern to protect the data of either reporters
or researchers from forced state examination. There is
little reason to believe that courts in the future will
sustain any better the protection of individual identities
and of the fiduciary relationship between informant and
researcher, reporter or other data-collector. When confidentiality conflicts with claims for national security,
claims by prosecuting or defense attorneys for data bearing
on legal actions, claims of need by law-enforcement personnel, demands by judges who wish to sustain the sovereignty
of the authority of the bench, and even the demands of
federal auditors, it is confidentiality that goes. In
1978 the Supreme Court made decisions which have raised
serious questions about the relative freedom that even
newspapers, with their clear First Amendment status, can
claim from police searches of their files and from
subpoenas to produce notes and data identifying their
sources of information. Given their power of publicity,
newspapers can defend themselves considerably better than
can social researchers and record-keeping institutions.
Thus, while I agree that an effective security system can
protect records against <u>un</u>authorized disclosure, the agents
of the state can easily gain the authority to force disclosure. And it is the state which has the most power to
create the most difficulty for the individuals so identified. Under such circumstances, researchers become merely
an investigatory arm of the state, collecting data under
the false pretense of independence.

My position is that the only true security system lies
in the destruction of identifiers, and that the independence
of social research can be sustained only if the routine
practice of identifier-destruction is adopted by researchers.
To my mind, in a time when records are proliferating and
the technical capacity to store and retrieve them is more
than keeping pace, and when both private and state investigative activities employing such records are increasing
at an alarming rate, the benefit of assuring personal
privacy more than outweighs the potential loss to social

research were identifiers to be destroyed. True, the
destruction of identifiers in records would prevent follow-
back studies, but it would not prevent longitudinal research
from taking place, nor would it prevent making fruitful
analyses of the data in aggregate form. Thus, while the
lack of identifiers makes the pursuit of certain kinds of
research questions more expensive and less convenient than
would be the case otherwise, it does not actually prevent
any basic kind of research. It does, on the other hand,
prevent the use of stored data to abuse Fourth and Fifth
Amendment rights and sustains the fiduciary relationship
between data-collectors and their informants.

But why should social researchers take it upon themselves to espouse a position which, albeit in only a small
way, is against some of their own interests? After all,
destruction of identifiers is so final. Were they preserved, some fascinating, perhaps intellectually critical
studies might be made. What do social researchers gain by
urging the destruction of data for which they themselves
might conceivably have some use? They would gain trust in
their independent status, and thus continue to gain access
to data which might very well be withheld from them if
they were agents of the state. And the state would gain
as well, for data obtained by independent researchers in
a position of trust must be seen as an empirically essential
corrective to the profoundly flawed data in official
records.

An enormous amount of information is now available in
the records maintained both by private interests and by the
state. These are the records to which Dr. Robins wishes
easier access than is now possible. But how reliable and
valid are the data in those records? I would suggest that
their validity and reliability are quite uncertain. People
fill out records in very special ways, attempting to shape
answers to serve their own ends. When they believe it is
in their own interest, and that they can get away with it,
they lie, they distort or they withhold data selectively.
Such distortion can run from giving themselves the benefit
of the doubt and neglecting to mention some things--as we
all do in our tax and employment application forms--to
total fabrication. Distortion by informants is compounded
by distortions and inaccuracies created by the record-
keepers themselves. No one can deny the weight of evidence
showing that police officers, teachers, doctors, social
workers, and other functionaries distort and even fabricate
some of what they themselves enter into the record in order
to justify their own actions and "cover" themselves. A

society which has to make decisions solely on the basis of
such data may make costly mistakes.

I would argue that our society needs data collected
independently of powerful state and private agencies in
order to be able to put into a realistic perspective the
distorted information to be found in the records maintained
by its major institutions and the distorted and self-
interested (even if well-intentioned) policy positions of
the officials of those institutions. The press is an essential source of such data, but so is social research.
Indeed, social research has potentially a very special
role, capable of obtaining better data than the press,
because it is not obliged to produce the who, what, where,
when and how of news. Social research can offer confidentiality of a sort that the reporter cannot promise.
Characteristically, when they report their findings social
researchers conceal identities not only of individual
informants, but also of those individuals they are concerned
with studying and of the places and institutions where the
actions and events they study occur. This gives social
research a potentially more powerful appeal by which to
encourage honest disclosure on the part of informants.

Reporters do not have a story if they do not in some
way identify a person, group of persons, institution or
community. In contrast, researchers have a story even when
all significant identifying details are withheld. When
concern is with illuminating the general characteristics
of some phenomenon, concrete identification is unimportant.
The social researcher may report his findings pseudonymously
from Yankee City or Middletown, but the reporter must write
his news from Newburyport or Muncie. When it comes to the
writing-up of findings, therefore, it is possible for social
researchers to offer a considerably greater degree of
anonymity to those they study than can journalists. But
what both can offer in good faith depends on how well they
can protect identifying materials. And the known degree
to which they have effectively protected identifying
materials seems likely to have a direct bearing on whether
those they study are open and honest, or whether they
provide the same self-protecting distortion that they
insert into official records. The value of what social
researchers can offer to society is thus directly connected
to the degree to which researchers are trusted as persons
standing apart from the state, attempting to describe and
understand and analyze the world for purposes of understanding rather than of law-enforcement.

Ordinarily, the agents of the state are indifferent to social research, and so give the illusion of safety, even privilege to identifying materials. This is no cause for being sanguine, however, for as I have already indicated, the state may not be counted on to support the protection of identities whenever it contravenes the purposes of its more powerful agents. It is then that social researchers get coopted or coerced into yielding up identifying data for investigative or prosecutorial purposes, even though the data were originally collected on the basis of an understanding that they would be used only for research purposes. Such breaches of faith, even if coerced rather than wilful, can neither sustain the trust of respondents nor the collection of fresh and accurate data which presupposes such trust. If, however, social researchers lobbied forcefully for a policy which required that in all records identifiers were to be destroyed wherever possible, even if at the expense of some kinds of research, and if they also adopted for themselves the policy of routine destruction of identifiers from their own data, then they could both sustain the trust of those they study and protect their own role of independence from the state. Should social researchers adopt such a policy, I believe, their data will be the richer and their moral and intellectual position the stronger.

17

Regulations of Research as Factors of Accountability and Power

Hans O. Mauksch

The issues of research regulations must be addressed on various levels. The basic question of the justification of research regulations provided the theme for earlier chapters. Evidently, the legitimacy, the need, and the ethics of regulating research remains an openly debated issue.

Regulations do exist. Questions are being raised about the consequences of the existence of such regulations. This could be construed as a narrowly focused, rather detail oriented concern with questions of adequacy or appropriateness and attempts to arrive at a judgement whether these are good regulations or not. In a more broadly conceived way, one can raise questions about the consequence of the very existence of regulations regardless of their judiciousness, fairness, and effectiveness.

I would like to raise some questions which I claim to be basic ones even though they are mired in detail. My colleague and former teacher, Albert J. Reiss, Jr., asserted quite correctly that regulations, as implemented, tend to protect the investigator rather than the subject. I should like to take a more austere point of view and claim that, there is a structural characteristic associated with bureaucratized rules which justifies the claim that regulations, by the very nature of bureaucratic procedures, end up protecting the regulatory process rather than either investigators or subjects.

Regulations become social institutions; they fight for continued existence rather than remaining subject to continuous scrutiny and critique. Regulations should always be accountable to two basic and over-arching questions: a) Accepting the fact that this regulation exists, does it actually accomplish what it purports to accomplish?   b) Regardless of its effectiveness in meeting its primary objectives, does the regulation incur unintended and potentially costly side effects?

It is these two questions which this paper wishes to raise. Let us use this is indeed so. Yet, the current process of obtaining voluntary informed consent tends to increase and accentuate the power relation which already exists. In lower-class experiences, a signature is associated with additional obligations and with giving away rights rather than assuming them. The hospitalized patient, who perceives the white-coated researcher to be a powerful figure, is likely to experience the consent form as a further instrument of power which can be rejected only at social costs and potential risks.

Conversely, the social sciences researcher who seeks to unobtrusively find acceptance in an alien social environment finds himself with little power. Indeed, the consent form gives power to those from who the researcher seeks acceptance.

The second modifier to the word consent is the word "informed." Here, too, the stereotype of a middle-class, literate community, makes achieving this goal questionable. I raise here again the question of the impersonal self-interest of any bureaucratic system which, in order to prevail, must find ways by which its regulations become universally applied.   The universalistic bend of regulations depersonalizes and ritualizes this information giving process in such a way that even the literate, middle-class subject can be confused and alienated.

In order to obtain a measure of the effectiveness of this aspect of the regulation, a study should ascertain the actual level of understanding achieved through formal consent forms as compared with interpretations provided by researchers sensitive to the needs of individual clients. There is evidence that personalized information experienced as a shared relationship is more effective than is a formal, impersonal, distance-suggesting communication structure. The impersonal, formal consent form with its legalized language of explanation is almost reminiscent of the adversary situation generic to the legal culture. The above suggested research should explore the effectiveness of information transmission as it is affected by its formalization.

The problem suggested here relates to a troubling but basic principle. No matter how worthy a social goal, centralizing its pursuit will give rise to consequences of large systems. A convincing case can be made for the argument that, as the center of controls and authority is located further away from the scene of application, implementation of the principles becomes more ritualized, routinized and less adaptable to the variations of local settings. The subtleties and depth underlying the crucial notion of "voluntary" and of "informed" consent should be the target of an important and fascinating case study. Between the conceptual debate about the merits, ethics, and need for social regulation and the controversies about the technicalities of the specifics of various regulations, lies the important existential question of whether the social need, if identified, can indeed be accomplished through a nationally controlled regulation.

Let us now turn to the second question which was suggested earlier in this presentation; "Are there unintended side effects?" The above suggested study of the actual working of this regulation might discover that the regulations which require the formalization of the research subjects' consent has at least one, surely unintended side effect. The formalization and legalization of this procedure serves in a subtle, but profound way to absolve the investigator of an internalized sense of responsibility and accountability once the consent form has been signed. Like all such statements, this proposition is extreme and surely subject to considerable individual variation. Yet, the point must be made that formalization and centralization must give the researcher a sense that he/she is less responsible individually since it is now a "signed contract."

Two alternative pictures can be suggested, admittedly, oversimplified and sketchy. A traditional model places responsibility for the welfare of the subject into the professional control systems surrounding the researcher. This includes presumed internalized norms, the informal surveillance of peer group and colleagues, and the personal agreement between researcher and subject. The other model substitutes for this local and informal accountability system, a formalized responsibility mechanism involving anonymous depersonalized, distant control. The former model had proven to be deficient and severe violations of individual responsibility have been recorded, at times at a shocking level. Yet, I question whether the remedy does not incur costs which are also serious. By succeeding in the reduction of the dramatic violations, we may have brought a system which erodes the professional responsibility and accountability systems as they are generally functioning.

Essentially, underlying these comments is the recognition that, no matter how desirable the social aim, no matter how carefully conceived the current policy and no matter how detailed the written form, the actual implementation is still within the domain of the researcher. The researcher still has the real power, the situational opportunities and the status to influence the consent giving process and to manipulate the understanding gained by the subject. What is written and what is signed is not necessarily a reflection of what really has taken place.

Let me turn to another example of an area where we could examine what I call the unintended consequences of a social regulation and procedure. Local review boards have been established to protect the interest of subjects on the local level presumably by mobilizing a group of peers who are to judge whether a given research protocol involves risks to subjects and, if so, how to assure informed voluntary participation. These local review boards supposedly are to judge the impact of the research project on the protection of human subjects but are not to judge the merit of the research question itself or the methodology used to conduct the research.

In fact, a careful study of local review boards might well reveal that these review boards serve functions other than those intended. I need to re-assert that I am not implying intentional distortion or misuse but rather that I am suggesting unintended, structural consequences, not within the deliberate or even conscious perspective of those involved. I propose that a study would reveal that local review boards vary significantly in the degree to which they enter the domain of control over research questions, topics, and methodology. I suggest that this is particularly true in review boards associated with institutions dominated by one profession and less so in the heterogenous environment of an Arts and Sciences college.

It is interesting that at a hearing before the National Commission for the Protection of Human Subjects the testimony by the American Sociological Association and by the American Nurses Association independently raised the issue of such problems associated with the clearance of research protocols through review boards in medical institutions in which the dominant membership of the local review board was comprised of physicians and biomedical scientists. Both testimonies suggested that these review boards, probably without conscious intent, judged research protocols by

individuals defined as not being part of "our" research
world more cautiously and more suspiciously. Questions of
appropriateness of research and of the validity of method-
ologies alien to biological scientists are frequently
raised by local review boards in such settings. Without
delving into the details, let me suggest that the kind of
reality assessment of policies and regulations which I have
outlined earlier would include the test of extraneous consequences of the creation of these review boards. The
question should be raised whether the existence of review
boards may serve in certain settings as an unintended but
actual arm of the wielders of power and status and thus,
serve as an instrument of social control quite incidental
to and actually in conflict with, the intent of the Commission for the Protection of Human Subjects.

In this presentation, I have suggested the need for a
realistic approach to the implementation of social goals
through policies and regulations. It occurs to me that
there is an available model which suggests itself for use
in testing the appropriateness of regulations, their effectiveness, their impact and their unintended consequences.
The last few years have seen the emergence of programs
called "Social Impact Analysis." The concepts and methodology of "Social Impact Analysis" arose in connection with
increasing concern with the environment and with the need
for the capability of assessing a broad range of consequences caused by the introduction of new technology in a
given setting. What is beginning to be an impressively
useful approach in the field of technology and natural
environment could well be developed to be applied to the
impact of new policies on social arrangements, programs,
individuals, and institutions. In a society in which every
addition to the physical world is carefully evaluated and
studied and in which every new drug is laboratory tested
for many years, it seems peculiar that new policies with
profound consequences and impact are unleased and implemented without careful testing, experimentation, and without
evaluation and monitoring.

As social scientists, we should look at these phenomena
from the point of view of social process. All scholars are
concerned and involved in the basic question about the
legitimacy of controls over research. The humanists can
make a unique contribution in helping us sort out the conflict of values. However, once it has been decided that a
regulation should exist, the social process of implementation becomes the proper concern of the social scientist. A
methodology which would seek to answer the two questions
suggested at the outset of this paper might offer a

major contribution to the delicate balance between freedom
and restraint in the field of inquiry by adding the
dimension of social feasibility and consequence.

Part V
Summary and Conclusion

18

Classification
of Some Problems

Dael Wolfle

Scientists find it useful to classify the phenomena with which they deal. Thus Professor Back mapped three areas of knowledge: the private, personal area that each individual tries to protect from invasion; a wider range that includes public information; and a remoter area of esoteric knowledge that may be forbidden for fear that possession would be dangerous. Professor Casper distinguished two types of controls over the freedom of research: restrictions imposed because the process of getting new knowledge involves the use of dangerous materials or dangerous activities; and restrictions imposed because the new knowledge itself is thought to be dangerous or because it may lead to harmful use. Dr. Hellegers accepted the first of these types, restrictions on research activities that threaten to be harmful, but otherwise accepted no limits other than the right of society to decide what research will be paid for from public funds.

Underlying all of these taxonomies or classificatory schemes one finds a conflict of values. The conflict is partly overt and deliberate. The search for new knowledge on which scientists place high value is given less value by many people, and is actively opposed by some. In fact Kurt Back has reminded us that freedom of inquiry has been an exceptional rather than a normal value in the history of society.

In addition to deliberate restrictions on the freedom of research, there are also constraints that are incidental to the achievement of other objectives. The Privacy Act was not designed to interfere with longitudinal studies, but the methods adopted to protect personal privacy have certainly had that effect.

Whether the constraints are deliberately imposed or are side effects of efforts to achieve other purposes, we must continue to plan to deal with conflicts of values. Our values as scientists are not universally shared; the processes by which we seek to extend knowledge are not always understood; there are other worthy and desirable objectives to be sought. Consequently, many decisions that will affect science will continue to be made by persons whose primary values are other than those that motivate the scientific enterprise. Moreover, after the basic decisions are made, detailed regulations will be written and compliance will be monitored by agency personnel whose primary interest is in the regulations themselves, and who are therefore quite likely to illustrate the bureaucratic sin of paying more attention to the details of the regulatory process than to the objectives that were intended to be served.

An example may be useful. The protection of human subjects is an objective most scientists would surely endorse. That obligation used to be left to the judgment of individual investigators. As time went on however, a few flagrant violations convinced the National Advisory Health Council that individual decisions no longer provided adequate protection. Accordingly, the Council adopted, and the Surgeon General promulgated, a simple one-paragraph statement that the National Institutes of Health would not support research involving human subjects unless the principal investigator's institution gave assurance that the research plans had been reviewed by a body of competent peers to make sure that proper procedures for securing informed consent would be employed; that the rights and welfare of the subjects would be protected; and that the potential benefits of the study outweighed the potential risks to the subjects.

This requirement was generally accepted and fairly easily put into effect. But to regulators there is something pretty unsatisfactory about a rule or principle that can be stated in half a dozen lines, and they quickly went to work to remedy that defect. They have succeeded altogether too well; the regulation has gotten longer and longer, more and more detailed, and with more and more specific rules, not about research procedures, but about the process by which research plans are reviewed. What started as a simple clear statement of collective institutional responsibility has become many pages of regulatory detail.

I cite this bit of history as an example of a control that was initiated not by outsiders but from within the biomedical community. Thus it supports statements of some of the earlier speakers that scientists accept the propriety of some controls over the freedom of inquiry.

At the same time, however, we have to protect the research enterprise, to support the values of research, and to oppose or seek to modify those restrictions that unduly interfere with the ability to secure useful research data. If we do not protect these values, no one else will.

In our protective role, we will have to recognize that value conflicts are often involved. We will sometimes have to compromise, and to accommodate other objectives. We will also have to keep watch over the implementation and detailed rule-making, for even if a principle is acceptable the processes used to enforce compliance may be harmful. In short, we will be engaged in a continuing political process. If we are to engage successfully in a political process we must use political tactics. Both legislative and administrative records show that a considerable number of scientists have learned that art, and practiced it effectively.

But there is another and prior task to perform in our efforts to protect the legitimate freedom of research, and one that should not be confused with the political aspects. That is the task of carefully analyzing the effects of proposed or existing restrictions. Are those restrictions merely minor irritations, or do they seriously limit ability to obtain new knowledge? Are the other values that are sought great enough to justify the restrictions? Can we employ alternative techniques to secure the same ends? We can carry out this task most effectively if we act as scientists and not as political activists, analyzing carefully so that we know what we want to defend, and how, and why -- all so we can then enter the political debate well informed, and well prepared.

It does no good, in this process, to equate free scientific inquiry with unrestricted testing of nuclear weapons, as Professor Casper seems to have done. Nor is it helpful to confuse peer review of individual research proposals with the decisions of representative government on the amount of federal support available for various areas of research, as Dr. Hellegers seems to have done.

In contrast, Dr. Robins' discussion of the effects of privacy regulations on longitudinal research is an exemplary analysis of how, in quite specific ways, particular regulations reduce the usefulness of some well-established research procedures. Moreover, it should be noted, her analysis led on to suggestions of how methods of collecting data could be modified to contribute positively to the protection of the privacy of the subjects whose histories were being examined.

That example is a good one on which to end. We accept
the existence and worthiness of values other than the freedom
of research, but the protection of that value is our responsibility. We must be willing to negotiate, and sometimes to
compromise, but we can also employ techniques and controls
that support such values as safety and privacy. But if we
are to do all of this effectively, we should start with the
careful, objective, and detailed analysis of how proposed or
existing restrictions affect the attainment of research
objectives.

19

Research Regulation, the Public, and Professional Organizations

Keith M. Wulff

In this chapter I will state what I think are some of the major questions which relate to regulation of scientific inquiry by the society, and suggest answers to some of these questions posed by the various authors of the preceeding chapters. In so doing, I will suggest that certain questions actually do not address the problems connected with regulation of research, that they are diversions which keep us from addressing the more fundamental problems which must be solved in order for essential scientific research in certain areas to continue.

The question of whether there should or should not be regulation of scientific inquiry is actually a red herring. For better or for worse, scientific inquiry is now regulated and will continue to be regulated, particularly where public money is involved. So we should be looking at specific regulations to see if they are accomplishing their purposes-- assuming those purposes are valid--or if they are only hindering the pursuit of knowledge. In the same vein, attaching regulations as impinging on our freedom of thought may or may not accomplish anything useful. As several authors have pointed out, the regulations are aimed not at thought but at actions. We do not--and cannot--have freedom of action when that action is dangerous to the public, or even when it seems to be dangerous. A more fruitful response to regulations which are thought to be unreasonably restrictive would be to show that the inquiry being regulated is in fact not dangerous to the public. Moreover, as Newburger stressed, the questions should be, What is the impact of the regulation on innovations? and, What is the value of an innovation to the public?

How Can the Public Find Out What
Research Should be Regulated?

A major question which must be dealt with is, How can the public find out what research should be regulated? Hellegers seems to suggest that environmental impact statements should be filed for research projects and, on the basis of such statements, the public would decide if various projects are detrimental to society, as well as if the projects are worth funding. But I wonder if such a process would achieve the desired results. How would the decisions be made concerning which projects would have to file impact statements? The bureaucratic fallback probably would be to require impact statements for <u>all</u> publicly funded projects, and that would create an administrative nightmare. More importantly, I do not think an environmental impact statement would be of any significant help in protecting a public which is without the technical expertise to evaluate the science involved.

Moreover, mandatory impact statements (or mandatory review of all research) could actually result in scientists becoming <u>less</u> sensitive to the possible side effects of their work. Although the chapters on recombinant DNA research all emphasize that it was the scientists working in the area who called attention to the possible need for regulation, some scientists may feel that once an impact statement is filed (or approval from an internal review board is obtained) they have no further obligation to the public. Our regulatory actions should be aimed at maintaining or enhancing scientists' feelings of responsibility for the ethical and practical implications of their work instead of creating another agency in charge of "values." I am not suggesting that requiring environmental impact statements would result in scientists becoming amoral. The point is that requiring impact statements may cause scientists to think less instead of more about the possible dangers of their work, since, in an adversary atmosphere, one tends to gloss over difficulties and not ask for trouble.

A second problem is that an impact statement, if meaningful, would stifle creativity. To be meaningful, an impact statement would have to be very specific, and experimenters would not be able to vary their methods or goals without filing a <u>new</u> impact statement. Clearly, that would severely limit both innovation and research productivity. Work would have to stop until the new statement was approved. On the other hand, an impact statement dealing only in generalities would be of negligible help in determining specific dangers to the public from a given project.

The suggestion that by reading impact statements the general public can decide which projects are detrimental to society and which are worth funding is wishful thinking. Certainly the public, through Congress, should take an active role in deciding what general areas of science it wants to support. But to ask the public to decide which specific research projects to fund is a mistake. Involving the public in that degree of detail would result in the kind of counter-productive controversies which have surrounded funding for abortion and projects dealing with evolution. That I do not think many people want. But far more important is the fact that today's most popular research projects are not the only ones that should be funded if we want to continue to develop our store of fundamental knowledge.

In the light of the public's incontestable right to decide what general areas of science it wants to support, I think Sinsheimer has raised the right question when he asked if we hold any values higher than the pursuit of knowledge. In looking at the broad areas in which its money is spent, the public may decide that it values other activities more than the acquisition of further knowledge in certain areas. These are broad questions of policy, and in a democracy, there is no question of the public's ultimate right to decide how its money will be spent.

If the public is to make enlightened decisions about which areas of science should receive the most support, the public must become better informed about science and about how funding decisions are made. And, since presently most scientists also know very little about the political process involved in deciding how federal funds will be spent, scientists must become better informed as well. Following the political discussions that surround funding for science can be discouraging, since sometimes it appears that scientific merit enters very little into the decisions. It is possible that the politicians know what the public wants, and that decisions made on scientific merit are not now among those things. Since the public should decide, regardless of what scientists may think of these decisions, it behooves the scientific community to make a concerted effort to educate both their elected representatives and the public at large. In particular, scientists must pay much more attention to their contacts with the media and to developing the ability and willingness to explain themselves and their work in lay terms.

That said, the problem of who decides what research needs to be regulated still remains. The answer can only be to continue to emphasize the importance of scientists being

responsible for pointing out areas that need regulation.
Self-regulation has not worked as well as one might hope in
the medical and legal professions; however, I do not think
that these professions are analogous to the scientific community. The people most critical of science have so far
been the scientists themselves. The various scientific
disciplines are also critical of each other. Scientists
must continue to insist on openness and critical evaluation
of each other's works and methods. I believe it is the openness in which science is performed and the critical evaluations by knowledgeable peers that are the public's best
protection. If science becomes less open and if scientists
concern themselves with protecting their image, then even a
review committee made up of nonscientists would not be able
to function effectively.

If it is up to scientists to report areas they think are
in need of regulation, how can the scientists who report be
assured of a hearing? Casper points out the problems that
these scientists might have in mobilizing the public to
support their position. It seems essential that organizations such as the AAAS and the various other professional
scientific organizations recognize that they have a duty to
help these scientists be heard. Developing a mechanism for
providing this help is a matter which should be high on the
agenda of every scientific organization. It is possible
that this mechanism would include instituting a formal investigation by members of the public and scientists from
disciplines other than those directly involved in a particular problem. Funding from the scientific community itself
must also somehow be provided.

Who Should Do the Regulating?

In general, scientists should regulate themselves. However, once it has been decided that a certain area of
research needs a more formal type of regulation, who
should do the regulating? Newburger has done a good job of
summarizing the pros and cons of various types of regulations.
I would like to emphasize again that the scientists doing the
research should not be encouraged to think that the ethical
questions in his or her research can simply be turned over to
some other person, panel, or agency.

The question of whether only scientists or both scientists and lay members oversee the work is not as important
as keeping the work open to view by all concerned people.
Familiarity with a procedure may dull one's sensitivity to
the possible dangers, so I am inclined to include lay members
on all review boards. They may be sensitive to concerns that

members of the discipline may overlook. This same sensitivity may be achieved by having a review board consisting of scientists from a wide range of disciplines. However, whether a review board is successful or not does not seem to depend so much on its makeup as on the reaction of the disciplines themselves to proposed review methods. Unless review is taken seriously by the scientific community, the matter of who is on the review board is of little consequence.

What is the Responsibility of Members of the Scientific Community?

The increase in the public's demand for accountability has put added emphasis on the scientist's responsibility to keep his or her house in order. I see two basic responsibilities which must be accepted by all members of the scientific community today. The first is that, in so far as possible, scientists should conduct themselves in such a manner that added restrictions are not imposed on them. The second is that they accept the responsibility to continue to scrutinize present (and future) regulations to see if they are helping more than harming either the public welfare or the scientific enterprise.

The various disciplines must continue to be involved in discussing the ethical implications of their work. As stated earlier, the scientific community can best point out the areas of research that need to be regulated. But a discipline will have a difficult time regulating itself if the members can not freely express their opinions because of pressure from peers. Open and continuing discussion of ethics should be seen as an important part of a discipline's activity. The professional organizations should insist that discussion of the ethical problems of the discipline be included in annual meetings and in the professional journals. In addition, these organizations should assume the responsibility of seeing that "whistle blowers" are treated fairly.

Now and in the future, discussion of ethical problems in research must be an integral part of a scientist's graduate education. At the graduate level, students are particularly interested in what is or is not ethical conduct. Ethical problems are real problems for graduate students, and it is thought by some that if a person is likely to engage in unethical conduct it will be in graduate school. Whether that is true or not, most scientists will probably never again have the time (or think they have the time) to deal with ethical questions as thoroughly as they can in graduate school.

Rather than advocating that a special ethics course be taught in every department, I suggest colloquia on ethical problems in the discipline and discussion of ethics as a part of learning the methods of the subject area. Finally, graduate students will learn best about ethics and its importance by working with professors and other scientists to whom the subject is important.

In summary, members of the scientific disciplines will not help themselves by reacting to proposed or existing regulation with cries of angry indignation and rejection. Instead, if proposed regulations are unnecessary or unworkable, at least some scientists must take the time to gather evidence and, based on this evidence, to oppose the regulations. However, regulations now in force should not be treated indifferently. The general public (or at least Congress) has seen a need for the regulations now in the law. Now is the time for the disciplines either to accept or convincingly oppose regulation, otherwise it may be impossible in the future for scientists in any discipline to convince the public that they can regulate themselves.